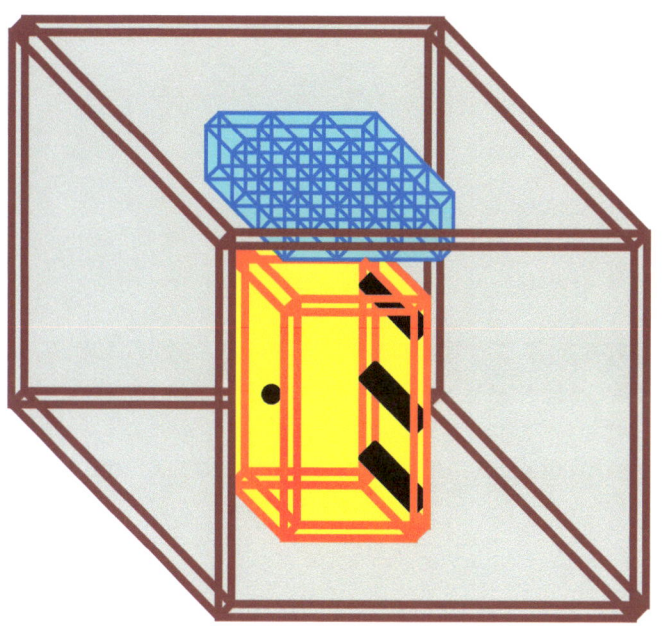

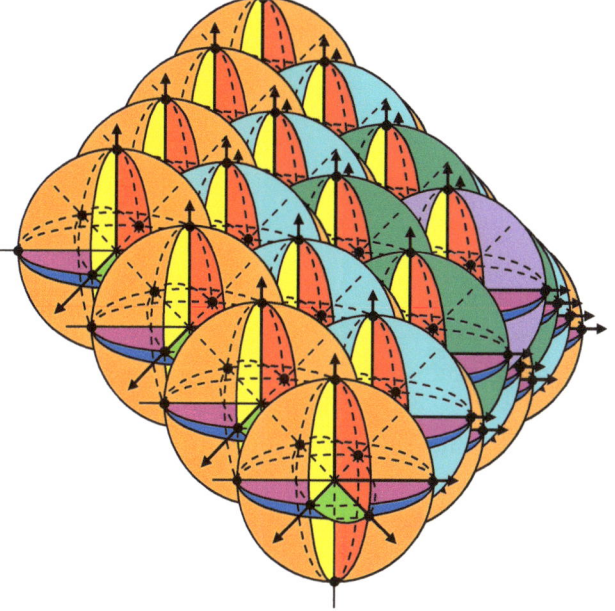

Full COLOR Illustrations of the

Fourth Dimension

Volume 2: Hypercube- and Hypersphere-Based Structures

Chris McMullen, Ph.D.

Full COLOR Illustrations of the Fourth Dimension, Volume 2:   Hypercube- and Hypersphere-Based Structures

Custom Books

Nonfiction / Science / Mathematics / Geometry & Topology

Nonfiction / Science / Mathematics / Recreation

ISBN: 1449529348

EAN-13: 9781449529345

Contents

Part I: Hypercubic 4D Objects (Tesseract- or Hypercuboid-Based Objects)

Basic building hyperblocks – putting tesseracts together   4

Hyperplane stacking – simple three-dimensional configurations of hyperblocks   5

Checkers in four-dimensional space with a three-dimensional checkerboard   6

Chess in the fourth dimension using a three-dimensional chess board   7

Sample variations of three-dimensional chess game designs   8

Hyperstorage – the simplest fully four-dimensional stacking of tesseracts   9

Rubik's tesseract – generalizing the Rubik's cube to four dimensions   10

Eight sets of 27 colored cubes on the hypersurface of Rubik's hypercube   11

Dissection of Rubik's tesseract – intersections of colored cubes on tesseracts   12

Rotating Rubik's hypercube through four-dimensional space   13

Hypercheckers – playing checkers with a four-dimensional checkerboard   14

Hyperchess – moving chess pieces through four dimensions   15

Sample variations of hyperchess and hypercheckers game designs   16

Great hyperpyramids – stacking tesseracts à la hyper-Egyptians   17

Hypercrosses – single, double, and triple crosses in the fourth dimension   18

Hyperfurniture – stacking tesseracts to make a hypertable and hyperchairs   19

Dinner for six four-dimensional beings – hyperchairs seated at a hypertable   20

Hyperstairs – building a cuboid-shaped staircase in the fourth dimension   21

Hyperdoors and hyperwindows – features of four-dimensional rooms   22

Living in four-dimensional space – a simple hyperhome   23

Architecture in the fourth dimension – sample hyperhouses   24

Reading hyperbooks – binding and turning hyperpages   25

A hypertelevision set – watching video in the fourth dimension   26

Hyperplane mirrors – reflections in higher-dimensional space   27

Part II: Hyperspherical 4D Objects (Glome-Based Objects)

The basic building hypersphere – dissection of the unit glome   28

Chains, planar arrays, and hyperplanar arrays of glomes   29

Simple hypercubic lattice – a basic four-dimensional array   30

Hypercrystals – simple lattice structures in hypercrystal solids   31

Hypersupermarket packing – stacking fruits at a grocery store   32

An astronomical coordinate system for four-dimensional space   33

Hypergeography – compass directions on the hypersurface of a glome   34

A hyperorbit – revolution of a spinning, titled planet about a star   35

Hyperletters and hypernumbers – a four-dimensional alphanumeric system   36

Hyperbilliards – a pool table designed for the fourth dimension   37

Hyperbowling – a four-dimensional bowling alley with hyperpins   38

Field hypersports – sample sports with a four-dimensional field   39

Hyperspherical mirrors – reflections from a metallic hyperbowl or hyperball   40

Part I: Hypercubic 4D Objects
(Tesseract- or Hypercuboid-Based Objects)

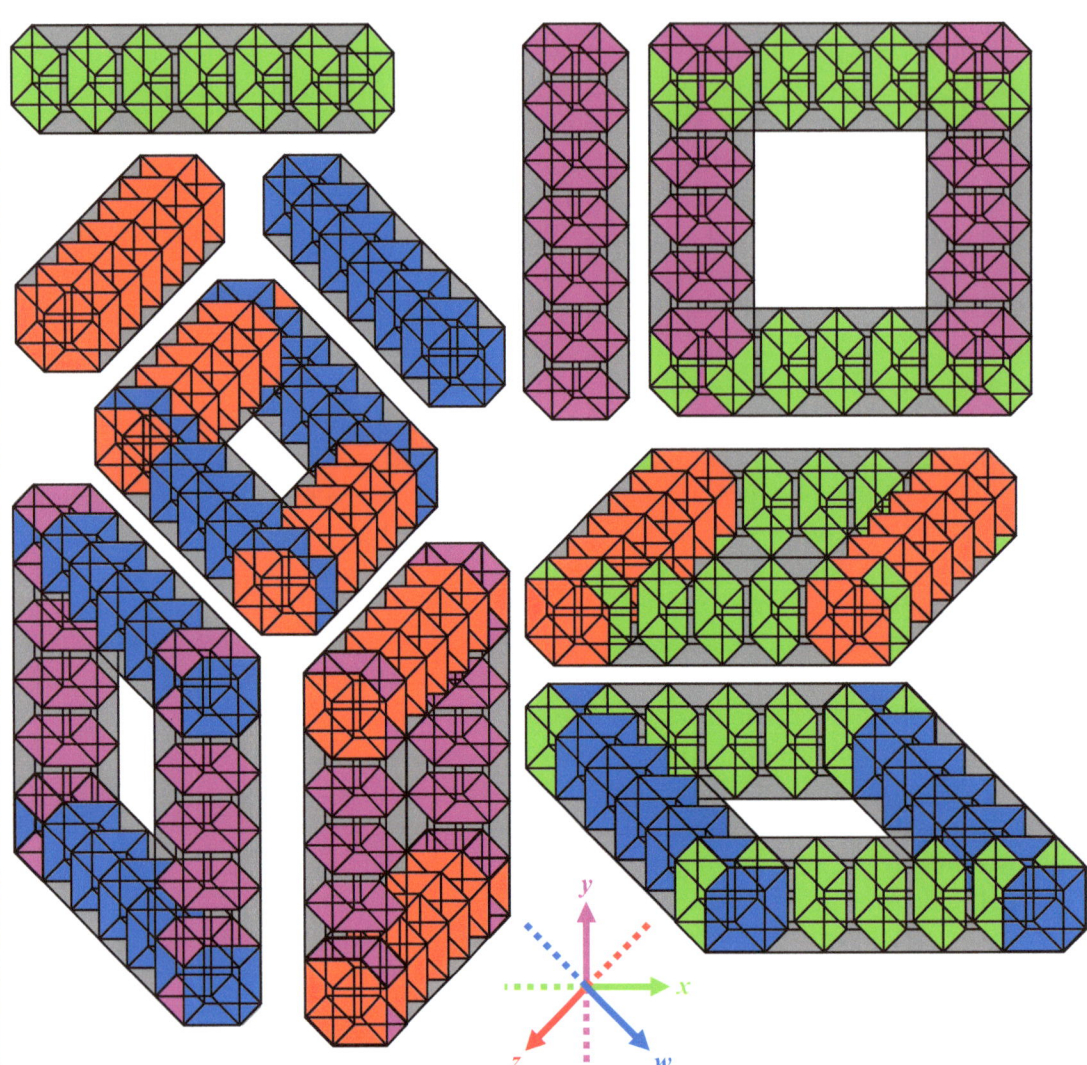

Tesseracts serve as the **building hyperblocks** from which any rectangular 4D object can be constructed. The four configurations at the top left show linear stacking of tesseracts in each of the four dimensions: left/right (x), up/down (y), front/back (z), and ana/kata (w). The terms **ana** and **kata** refer to motion back and forth along the fourth dimension. The remaining diagrams show square configurations in the six mutually orthogonal planes: xy, zx, wx, yz, wy, and zw (but not in this same order). Just as two stacked cubes meet at one square face, stacked tesseracts meet at one of their eight bounding cubes. The shared cubes are highlighted in the images above; the remainder of each tesseract is shaded gray.

These tesseracts are stacked in **three-dimensional configurations**, forming finite portions of **hyperplanes**. Each diagram above is a 3×3×3 stack of tesseracts in one of the four fundamental orthogonal hyperplanes – xyz, yzw, zwx, and wxy. Each stack consists of 27 tesseracts – 3 tesseracts in a row along each of the 3 dimensions used. In each case, 14 of the tesseracts are colored, 13 are blank to help visualize the hyperplane. For convenience, all of the edges are shown (even if they would be blocked by colored tesseracts). Such hyperplane stacking can be drawn by forming a row of tesseracts, then copying this row and placing the rows together, then copying this planar configuration and placing the planar configurations together (as shown in the small diagrams).

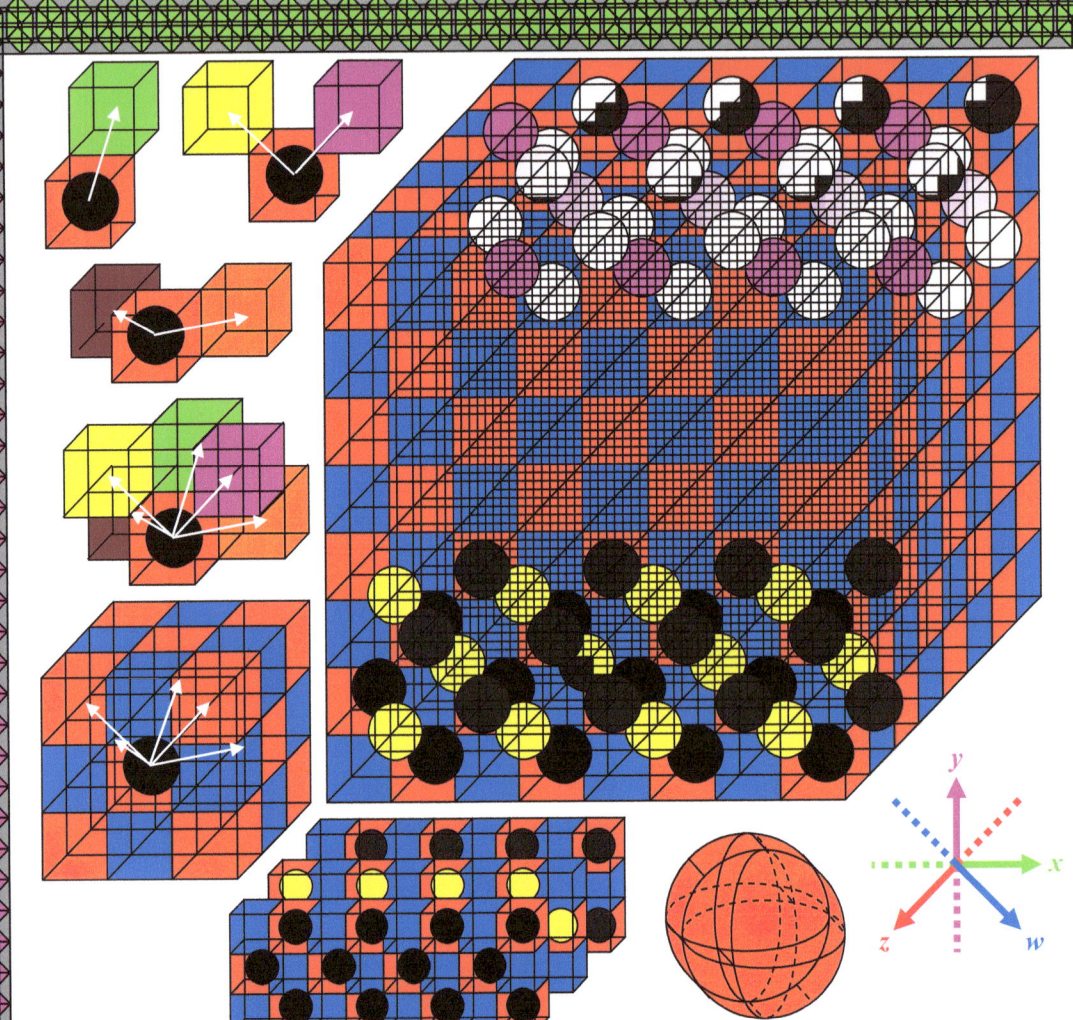

Four-dimensional beings could play **hypercheckers** on a three-dimensional **hypercheckerboard**, which could lie "flat" on the three-dimensional top of a hypertable (the board having very little hyperthickness). The hypercheckers would have the shape of spherical hyperdisks – spherinders (hypercylinders with spherical cross section) with just a little hyperthickness (along the axis of the hypercylinder). They would look like three-dimensional marbles, with a little thickness along the fourth dimension; but they would lie flat – not roll. The spherically shaped hypercheckers would lie "flat" against the three-dimensional cubic hyperchecker spaces (these cubes are solid throughout, just like the squares of a two-dimensional checkerboard are solid), since a cube and a sphere can be placed side by side and touch at every point on the sphere (inside and out!) in the fourth dimension, just as a circle can lie adjacent to a square in three dimensions (but not two). The 8×8×8 hypercheckerboard would have **512 spaces** (256 usable). If each color receives three rows and three hyperrows, each team would begin with 36 pieces. An unobstructed, unkinged hyperchecker not at the edge could advance in one of the **5 directions** shown (to any cube sharing an edge; cf. any square sharing a corner in two dimensions).

Four-dimensional beings could play **hyperchess** with a (three-dimensional) 8×8×8 **hypercheckerboard**. A smaller size board may be more manageable, since two-dimensional chess with an 8×8 board is already very complicated; however, hyperbrains could potentially hold many more neurons. A simple way to generalize 8×8 chess is to have three rows of pawns protecting the 8 major pieces; this is not the only way to generalize the chess board to three dimensions (and three-dimensional chess has been played in our three-dimensional world, if not from the perspective of the fourth dimension). Each piece would stand on a spherical bottom that rests snugly against any of the cubic spaces. The three dimensions of the board would be forward/backward, right/left, and ana/kata, while the pieces would extend upward (opposite to gravity). The kings' cubes have the same color. Here is a naïve version of three-dimensional chess: the king could move one space in any of 26 directions, since any cube not at the edge touches 26 other cubes; pawns could move forward or ana one space (or two spaces on their first move), or could capture one space diagonally like a hyperchecker (including *en passant*); rooks could move in one of 6 directions (right or left along a rank, forward or backward along a file, or ana or kata along a hyperfile – directions where adjacent cubes share squares); bishops could move in one of 12 directions, like kinged checkers (directions where cubes share edges); knights could move two spaces like a rook and one more space like a rook at a right angle to the first direction; and queens may be restricted to the 18 directions combining rook and bishop moves (omitting the 8 corner-shared directions).

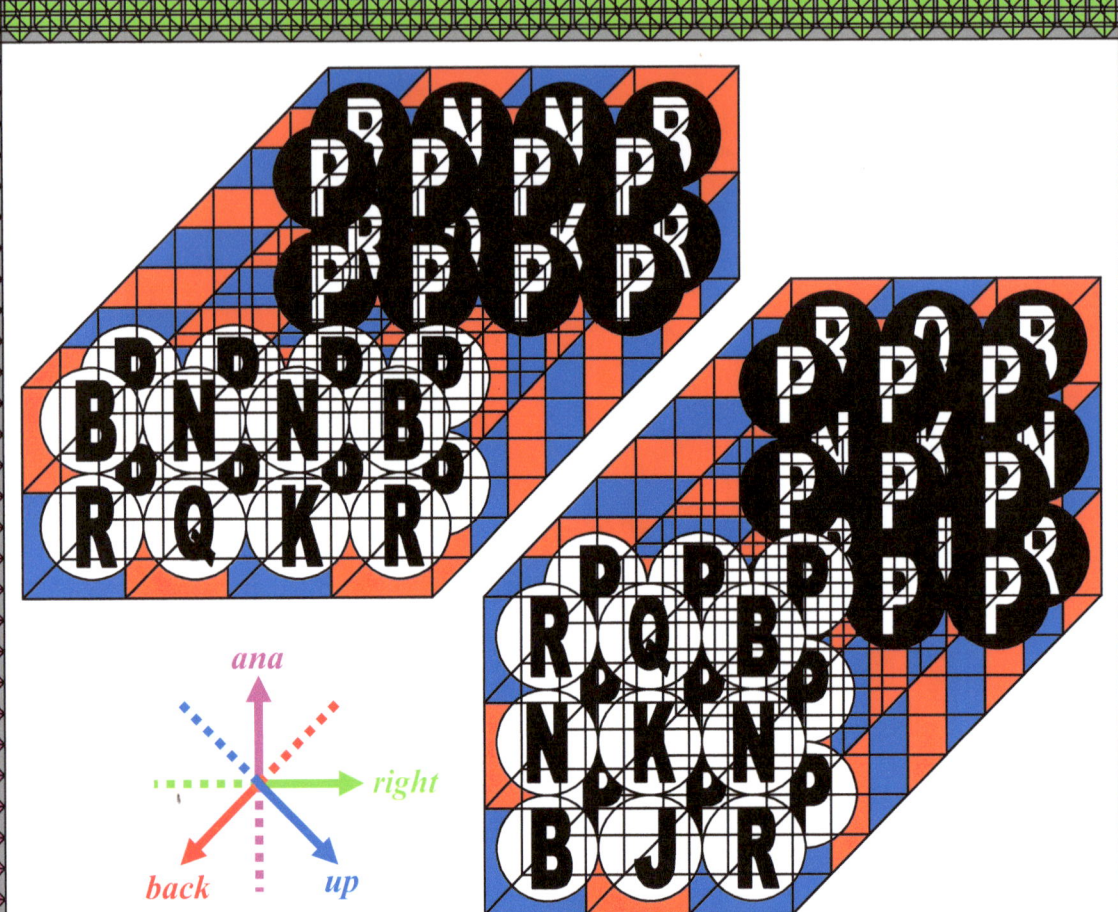

One interesting feature of playing **three-dimensional chess** on a 4×2×8 chess board is that it consists of 64 cubic spaces – the same number as ordinary two-dimensional chess on an 8×8 chess board. Also, each color begins with the same pieces, and the two armies are separated by the usual four ranks. Even so, the game would be considerably different. It would be fair to let the pawns advance straight toward the corresponding piece in the enemy camp only (except, of course, for diagonal attacks; but there the designer has a choice – do you allow diagonal attacks vertically as well as across?). The other pieces are more restricted on this three-dimensional game: Diagonal bishop and queen moves can only move over four files at most, for example. Pieces can move vertically, but only column over. On the other hand, the third dimension provides greater visualization challenges – knight moves would be a little trickier, for example. If castling is preserved, it would not be quite the same as two-dimensional chess. There are twice as many center cubes to control compared to the usual number of central squares; similarly, there are twice as many edge squares. The 3×3×8 game shown is similar, with 72 spaces, 3 ranks, 9 files, and 3 hyperranks (cf. hyperfiles on page 7). There is room for an extra piece, such as the court jester shown – which is confined to moving like a king, except that it may not leave the 9 castle squares, cannot attack, and cannot be killed.

These tesseracts are stacked in **four-dimensional configurations**, forming a finite portion of a **four-dimensional hyperplane** (cf. the three-dimensional stacking on page 5). The diagram above is a 3×3×3×3 hyperstack of tesseracts in four-dimensional space. This hyperstack consists of 81 tesseracts – 3 tesseracts in a row along each of the 4 dimensions used. Such four-dimensional hyperplane stacking can be drawn by forming a row of tesseracts, then copying this row and placing the rows together, then copying this planar configuration and placing the planar configurations together, and then copying this three-dimensional configuration and placing the three-dimensional configurations together (as shown).

Rubik's tesseract is a 3×3×3×3 (four-dimensional) configuration of tesseracts (81 in all). Rubik's tesseract is bounded by **8 colored "sides,"** where each "side" is a 3×3×3 (three-dimensional) configuration of tesseracts. Thus, there are 8 × 27 = **216 colored cubes** on the hypersurface (8 sets of 27 on each three-dimensional "side"). Compare to the 3×3×3 Rubik's cube, which consists of 27 cubes (26 cubes on its surface) and has 6 colored sides, each of which consists of 9 colored squares. The colored cubes in the diagram above are only partially colored in order to make the interior and rear visible.

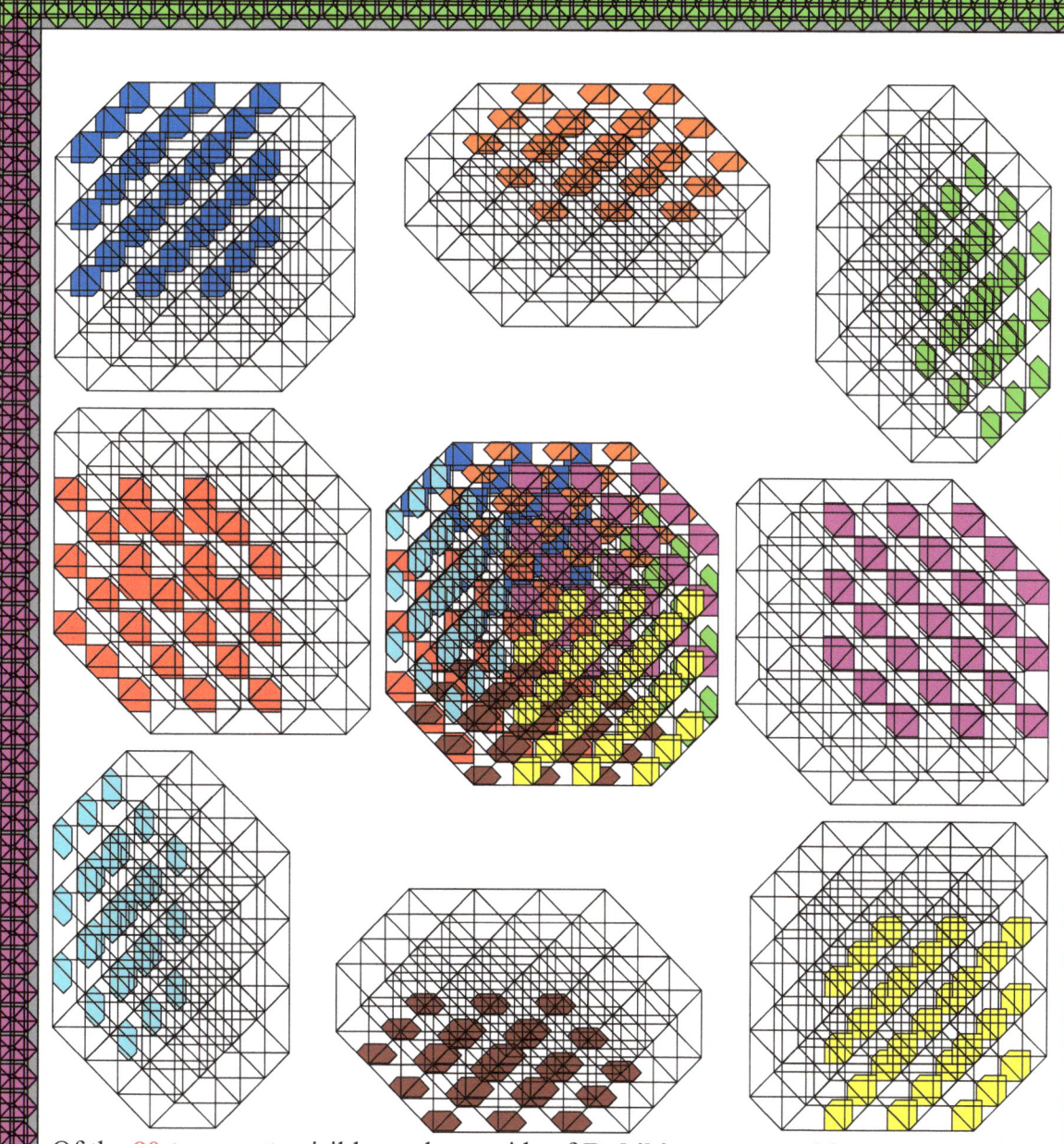

Of the **80 tesseracts** visible on the outside of **Rubik's tesseract**, **16** are **corners**, **8** are **body-centered** (lying in the center of a 3×3×3 "side"), **24** are **face-centered** (lying in the center of a 3×3 array, but not the center of a 3×3×3 "side"), and **32** are **edge-centered** (lying in the center of a row of 3 tesseracts, but not the center of a 3×3 array). The 81st tesseract lying in the geometric center is said to be content-centered; it is not visible from the outside. The 80 outer tesseracts can be grouped into the 8 sets of colored 3×3×3 "sides" illustrated above. Compare to the Rubik's cube, where the 26 outer cubes can be grouped into 6 sets of colored 3×3 sides – with 8 lying at corners, 6 at the centers of faces, and 12 at the centers of edges.

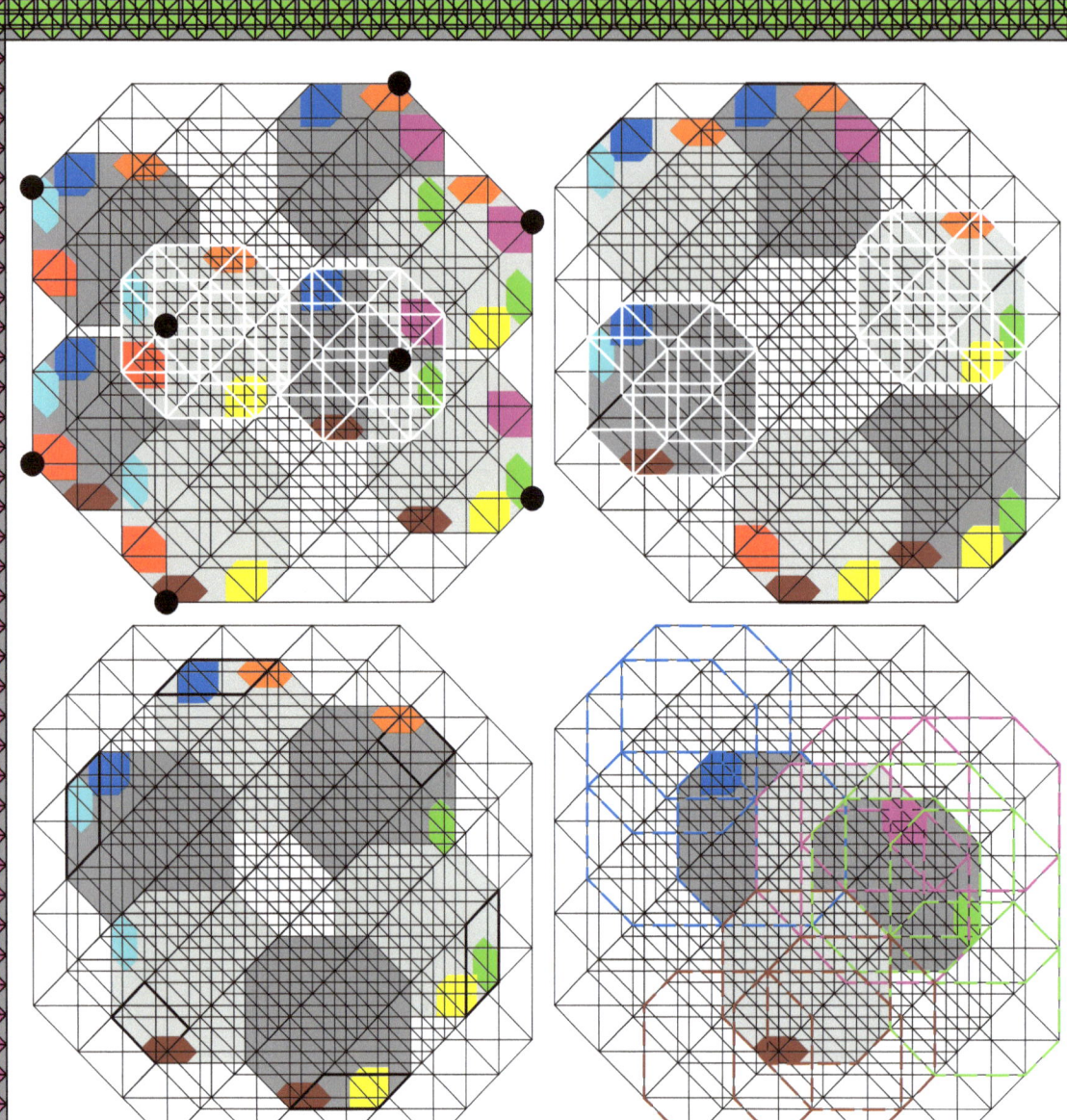

Of the **216 colored cubes** visible on the hypersurface of **Rubik's tesseract**, **64** meet at the **outer corners** of the 16 outer corner tesseracts, **8** lie in the 8 **body-centered** tesseracts, **48** share a common square in the 24 **face-centered** tesseracts, and **96** share a common edge in the 32 **edge-centered** tesseracts. 4 different colored cubes meet at each of the 16 outer corners, 3 different colored cubes meet at each of the 32 edge-centered tesseracts, 2 different colored cubes meet at each of the 24 face-centered tesseracts, and 1 colored cube resides in each of the 8 body-centered tesseracts. Compare to the Rubik's cube, which has 54 colored squares, where 3 colors meet at each of the 8 corners (24 squares all together meet at corners), 2 colors meet at each of the 12 edges (24 squares are edge-centered), and 1 color lies at the center of each of the 6 sides (6 squares are face-centered).

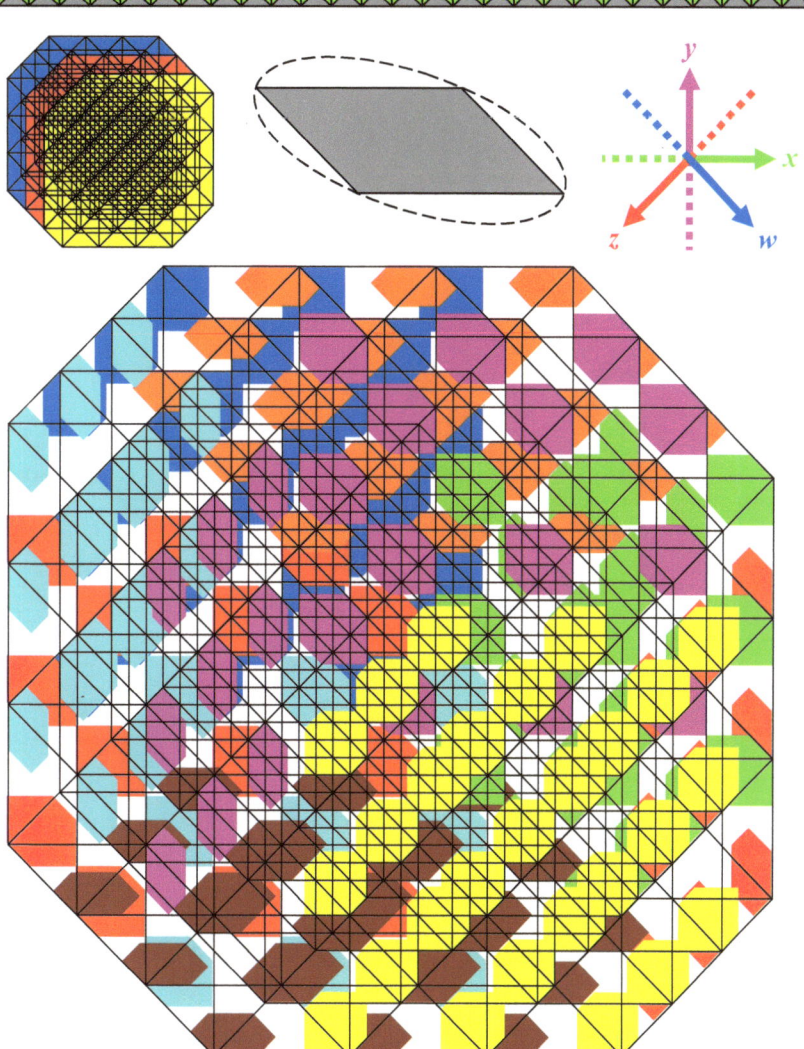

There are **6 distinct types of rotations** that can be applied to **Rubik's tesseract** – one corresponding to each of the 6 mutually orthogonal planes (xy, zx, wx, yz, wy, and zw). Through a single turn of Rubik's tesseract, 27 of the tesseracts on a common row rotate 90º with respect to the other 54 tesseracts. In the diagram above, the 27 positive w tesseracts rotate about the wy plane; the other 54 tesseracts remain stationary. The blue cubes are stationary. The yellow cubes rotate about their central vertical axis, rotating into a congruent position in 90º. The 9 positive w brown cubes and similarly orange likewise rotate into the same position in 90º. Look carefully and contrast with page 10: The positive w red wxy cubes rotate into turquoise yzw cubes, positive w turquoise rotates into pink wxy, positive w pink rotates into green yzw, and positive w green rotates into red. Compare to Rubik's cube, where there are 3 distinct types of rotations (lying in the xy, zx, and yz planes), any of which rotates 9 cubes on a given row 90º with respect to the other 18 cubes.

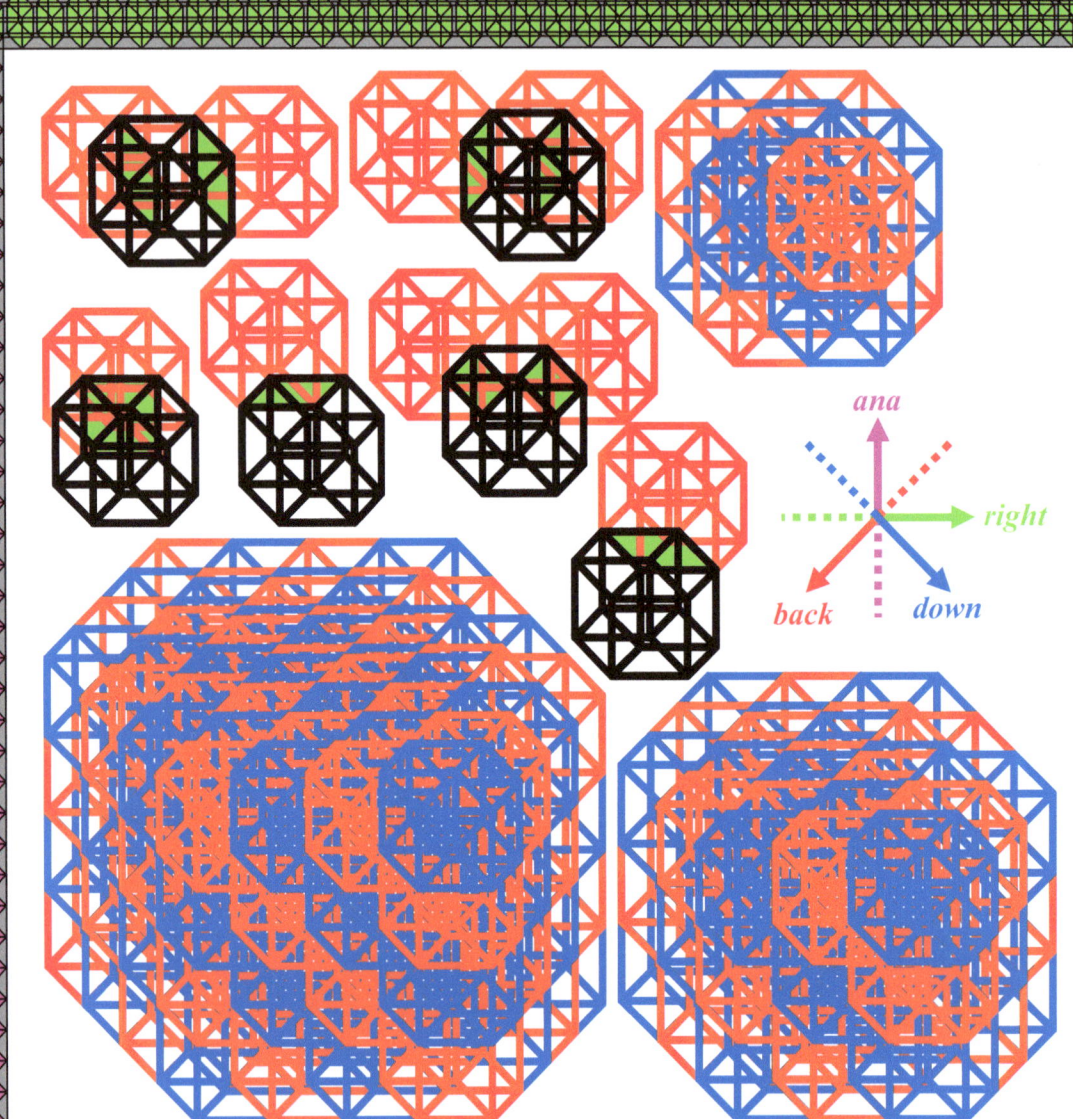

A conventional two-dimensional checkerboard generalized to a **fully four-dimensional hypercheckerboard** in the most straightforward way would be 8×8×8×8. It would consist of **4096 tesseract-shaped spaces**, half of which would be usable during a game of hypercheckers. If each color receives three rows in each of three dimensions (forward, upward, and "anaward" – with each row consisting of four pieces checked left/right along the width), each team would begin with 108 pieces. A smaller size board may be rich enough: For example, a 4×4×4×4 board would have 256 spaces (compare to the traditional 8×8 board, which has 64 spaces); in this case, though, the starting pieces cannot fill as many rows forward, upward, and anaward. The three boards illustrated in the diagram above are 2×2×2×2, 3×3×3×3, and 4×4×4×4. An unobstructed, unkinged hyperchecker not at the edge could advance in one of the **9 directions** shown (to any tesseract sharing a plane; cf. any square sharing a corner in two dimensions and any cube sharing an edge in three dimensions).

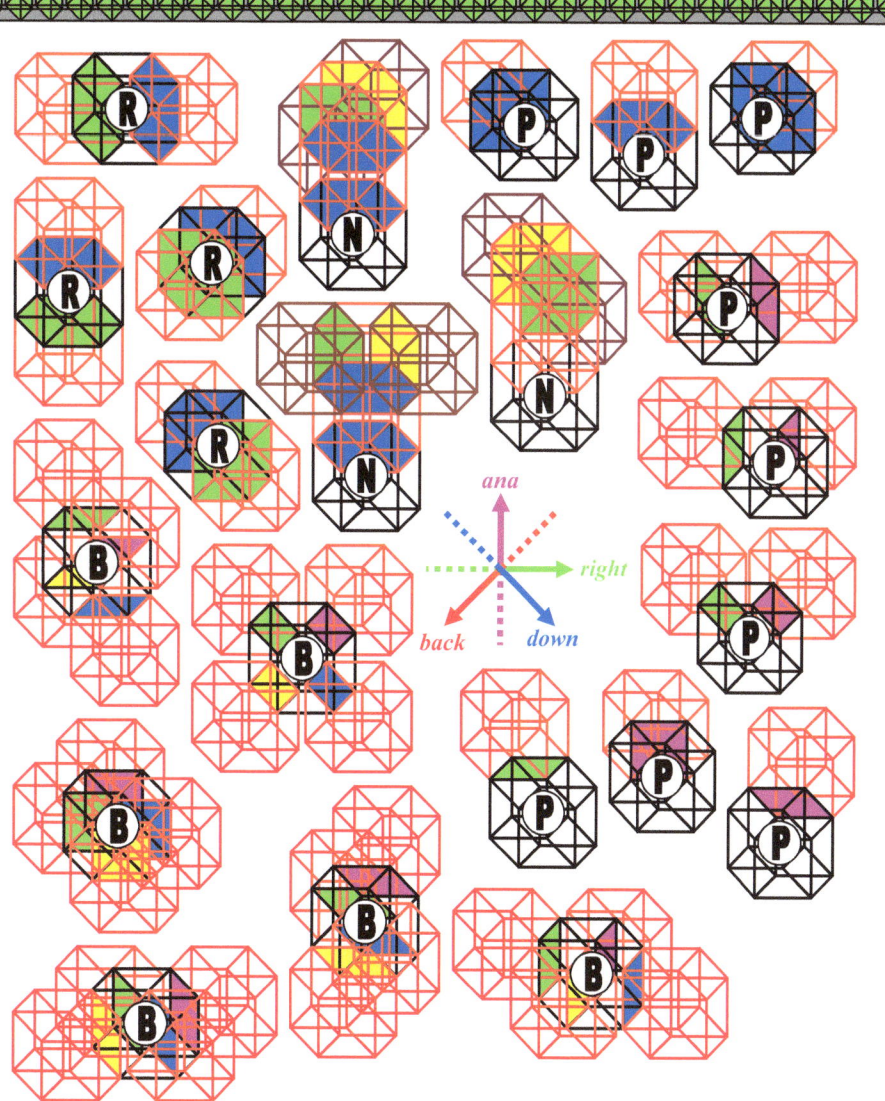

On a fully **four-dimensional chess** board, a king could move one space in any of 80 possible directions (since a king sitting in the center of a 3×3×3×3 grid connects to 80 tesseract-shaped spaces). A rook could move in 8 possible directions: right/left along a rank, forward/backward along a file, or up/down or ana/kata along two orthogonal types of hyperfiles. A knight moves forward two spaces like a rook, and then moves one more space perpendicular to the first direction (six choices for each possible starting direction). The straightforward analog of a two-dimensional bishop involves 24 possibilities – "diagonally" toward any adjacent tesseract sharing a square (in three dimensions, diagonal cubes share edges, while in two dimensions diagonal squares share corners). A queen that moves like a rook or bishop can then move in 32 different directions. Pawns advance along files or hyperfiles, or attack across one of 9 diagonals like hypercheckers – the number of bishop moves that are not backward, downward, or kataward.

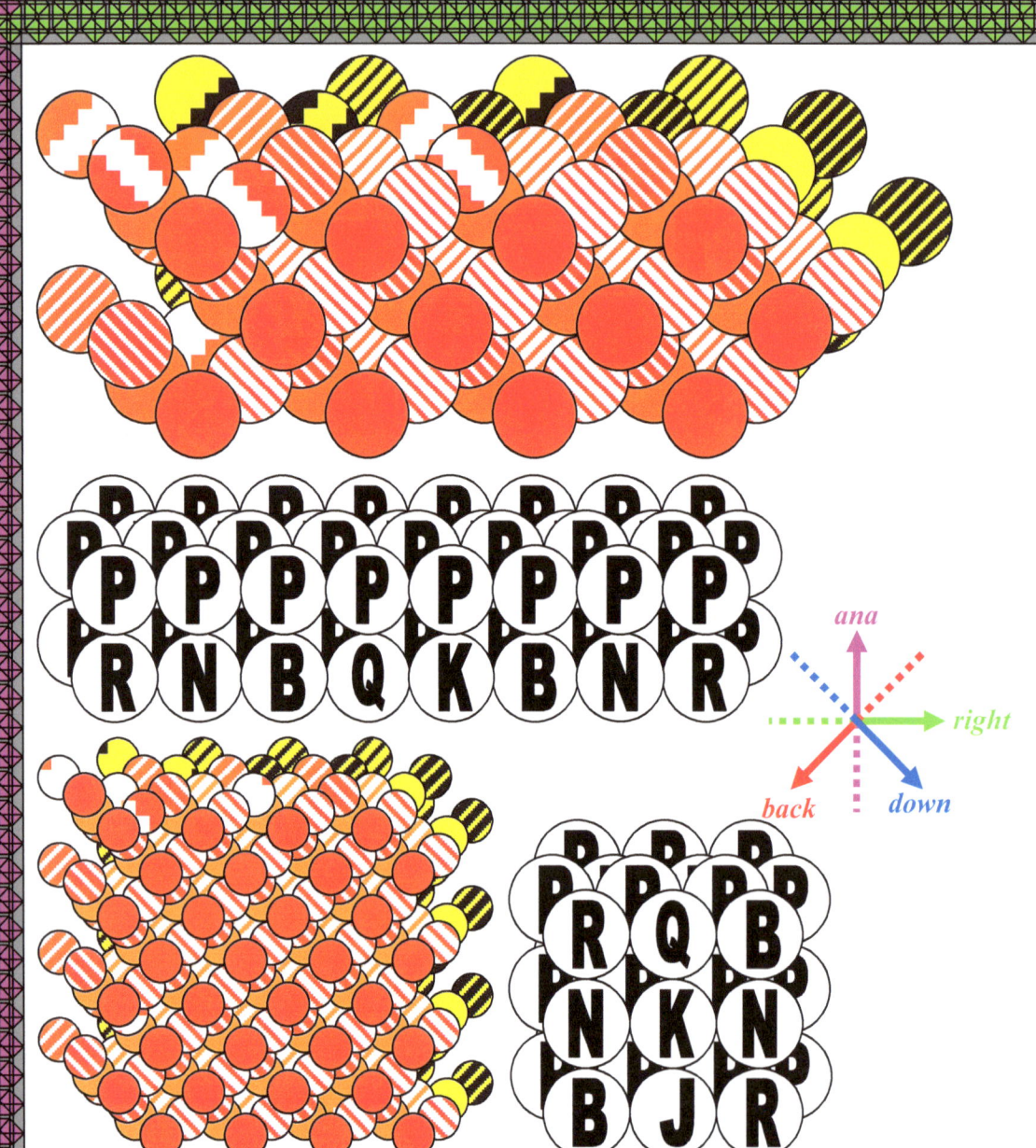

The top two diagrams show **hypercheckers** and **hyperchess** pieces based on extending the standard three-row and two-row piece arrangements along the upward and anaward dimensions; the hyperchecker arrangement consists of 108 hypercheckers and the hyperchess arrangement consists of 64 pieces (both counts are per army), 56 of which are pawns. The bottom two illustrations show that these games can instead by designed based on planar, rather than linear, configurations; this hyperchecker pattern features three sets of planar configurations along each of the two extra dimensions (288 hypercheckers per army), while this (smaller) hyperchess pattern has its major pieces arranged in a plane and protected by three sets of planar pawn arrangements – one along each extra dimension and their diagonal (36 pieces per army).

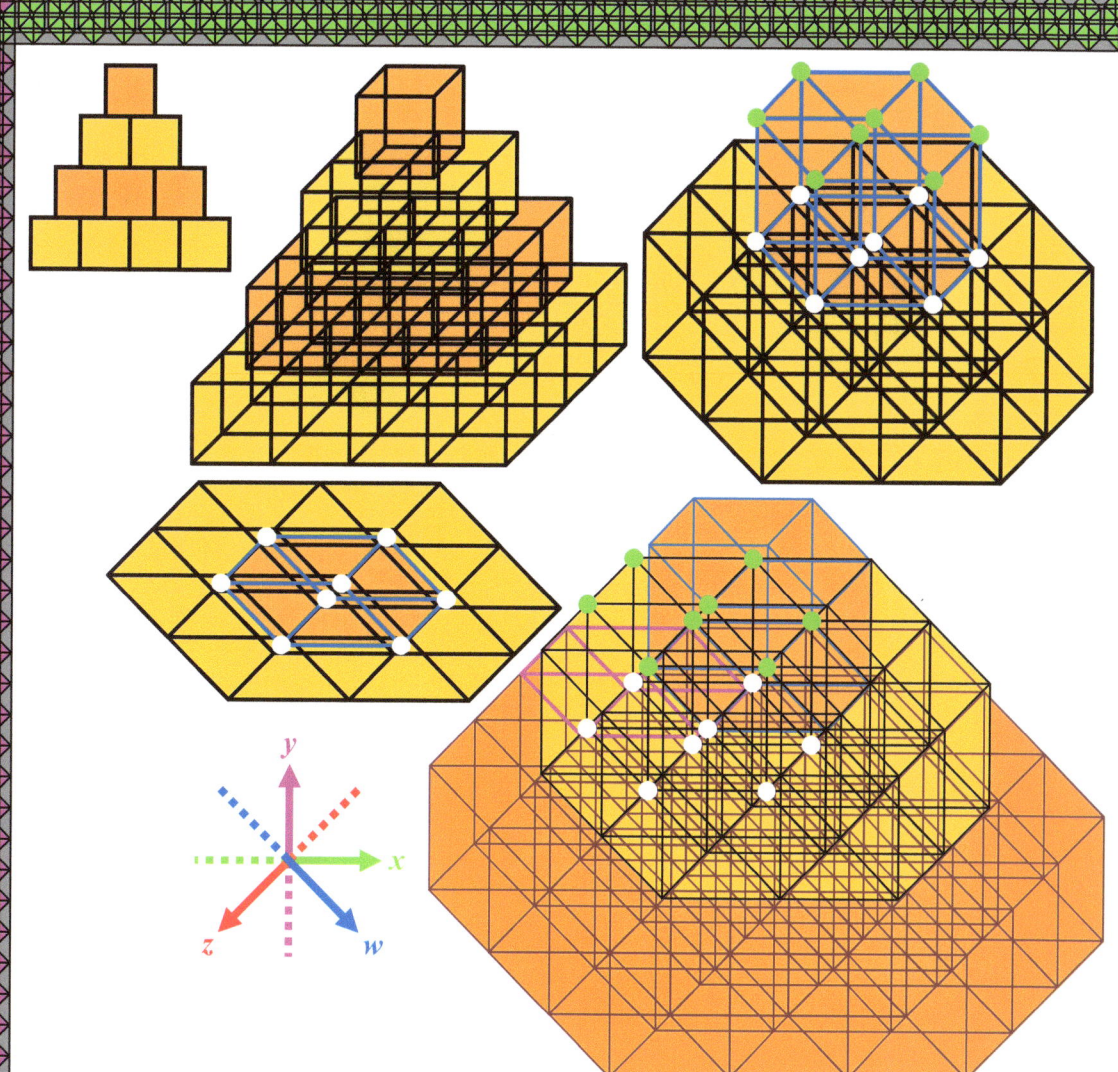

Three layers of a **hyperpyramid** built by stacking tesseracts are shown on the bottom right; the top right shows the top two layers. The top layer has 1 tesseract, the second layer has 8 tesseracts, the third layer has 27 tesseracts, and the pattern continues as the cube of the layer number (cf. cube stacking that goes as the square or square stacking that equals the layer number in three- and two-dimensional pyramids). The bottom 8 corners of each tesseract lie at the center of the 8 *zwx* cubes of the 8 tesseracts beneath it, where *y* is vertically upward in these diagrams (cf. the bottom 4 corners of a cube lying in the centers of the 4 top squares of the 4 cubes beneath it or the bottom 2 corners of a square lying in the center of the 2 top lines of the 2 squares beneath it in three- and two-dimensional pyramid formations, respectively). One diagram shows the top 4 cubes of the second layer of the hyperpyramid and how it relates to the bottom cube of the tesseract from the top layer. Another point to consider from a hyperengineering perspective, aside from the number of hyperblocks immediately below another, is that the mass of each hyperblock is proportional to $(\text{length})^4$.

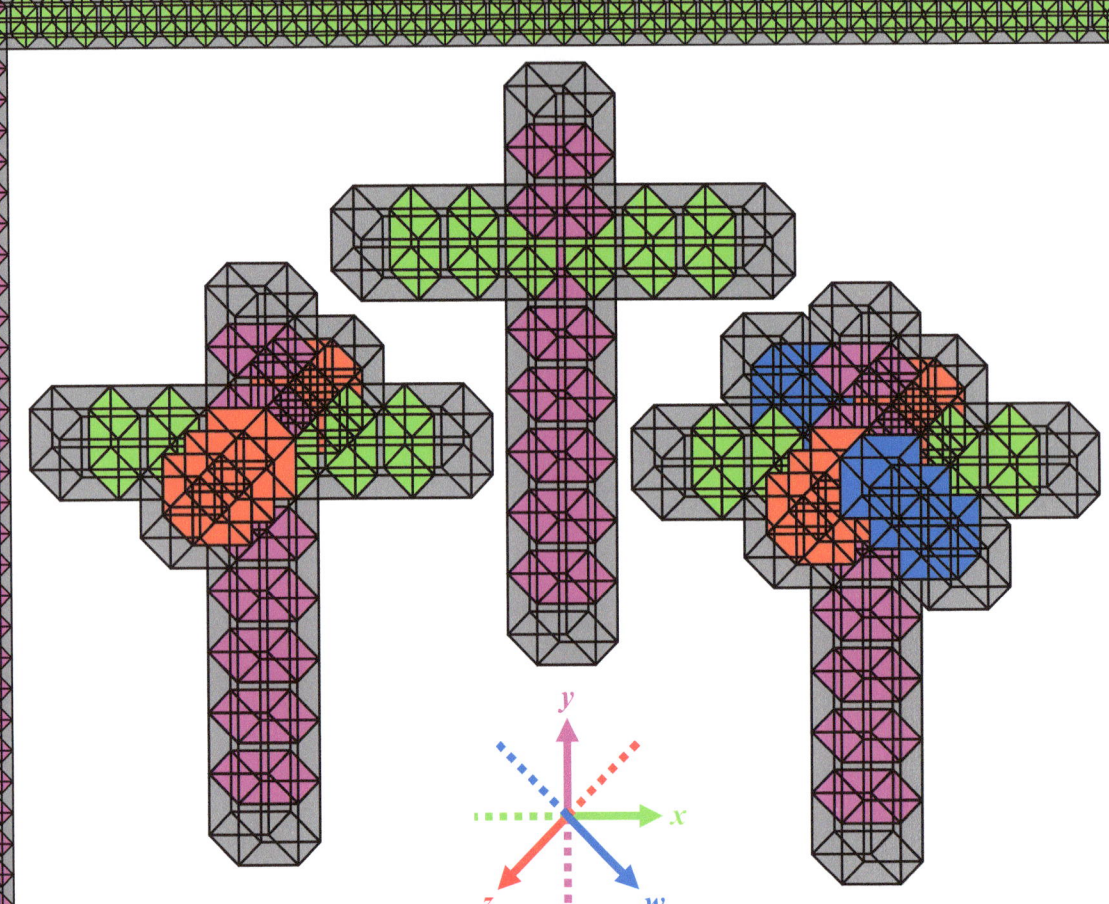

A four-dimensional **hypercross** can be a **single cross** (center), **doublecross** (left), or a **triplecross** (right). The colored cubes in the diagrams above are shared by adjacent tesseracts. A four-dimensional humanoid may have four arms – standing vertically, facing forward, this would be convenient for grabbing objects on the left, right, ana, or kata. If so, the double hypercross would be most suitable for nailing a hyperhumanoid to a cross in a fully four-dimensional universe. If Jesus Christ, or his hyper-equivalent, were to live in such a four-dimensional universe, **hyperChristians** would mount **hypercrucifixes** on walls and doors and attach them to necklaces. Such a four-dimensional humanoid would fit on the double hypercross just like a three-dimensional human can fit on a (single) cross: Observe that the fourth dimension (ana/kata) in the diagram on the left is perpendicular to the three large dimensions of the hypercross (which also has a little hyperthickness). Furthermore, such a hypercrucifix would lie flat against a wall in a four-dimensional home; the wall would have three large dimensions and a small hyperthickness. By analogy, consider a two-dimensional cross in a two-dimensional world, where it seems very strange indeed to hear the notion that such a two-dimensional cross could lie flat against a wall in three-dimensional space. This is because it is difficult to imagine rotating the cross 90º through the third dimension.

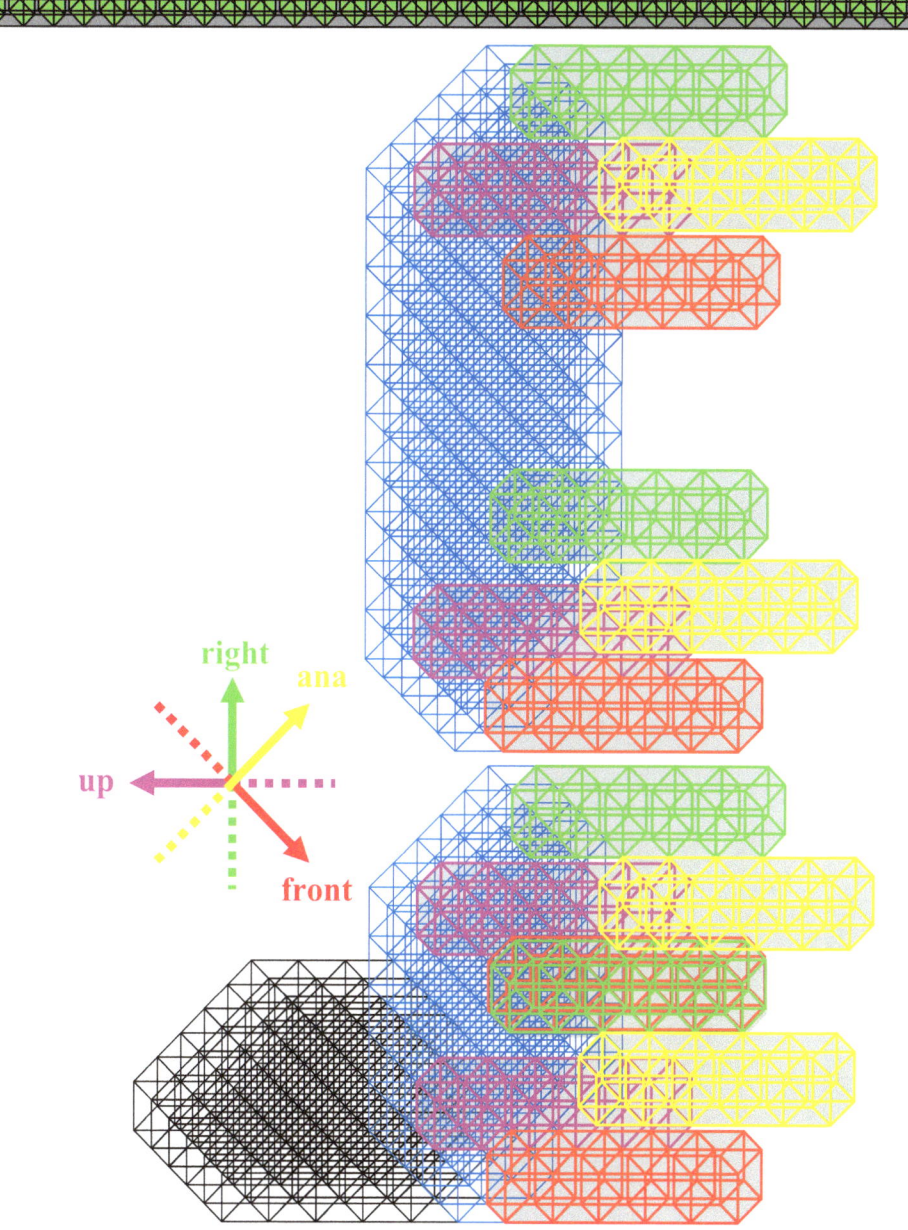

This **hypertable** (viewed sideways) has a tabletop with three large dimensions and a tiny thickness: It is 10 units wide (left/right), 5 units deep (front/back), 5 units hyperdeep (ana/kata), and 1 unit thick (up/down). A hyperhumanoid could sit at one of the 6 planar "edges" (left, right, ana, kata, front, and back); the top and bottom hypersurfaces are cuboids (three-dimensional rectangular boxes). The 8 legs are all parallel to one another, yet perpendicular to the each of the three large dimensions of the tabletop (the colored tesseracts comprising the legs are shown in front in order to make them more readily identifiable). The **hyperchair** shown is similar in design, except that it also has a back with three large dimensions – height, depth, and hyperdepth – and a tiny width.

This illustration shows six **hyperchairs** in relation to the six (three-dimensional) cuboid-shaped "sides" of the **hypertable** (which is bounded by eight cuboids; the other two constitute the top and bottom "sides" – a cuboid is a rectangular box, different from a cube in that the lengths of different types of edges may differ).

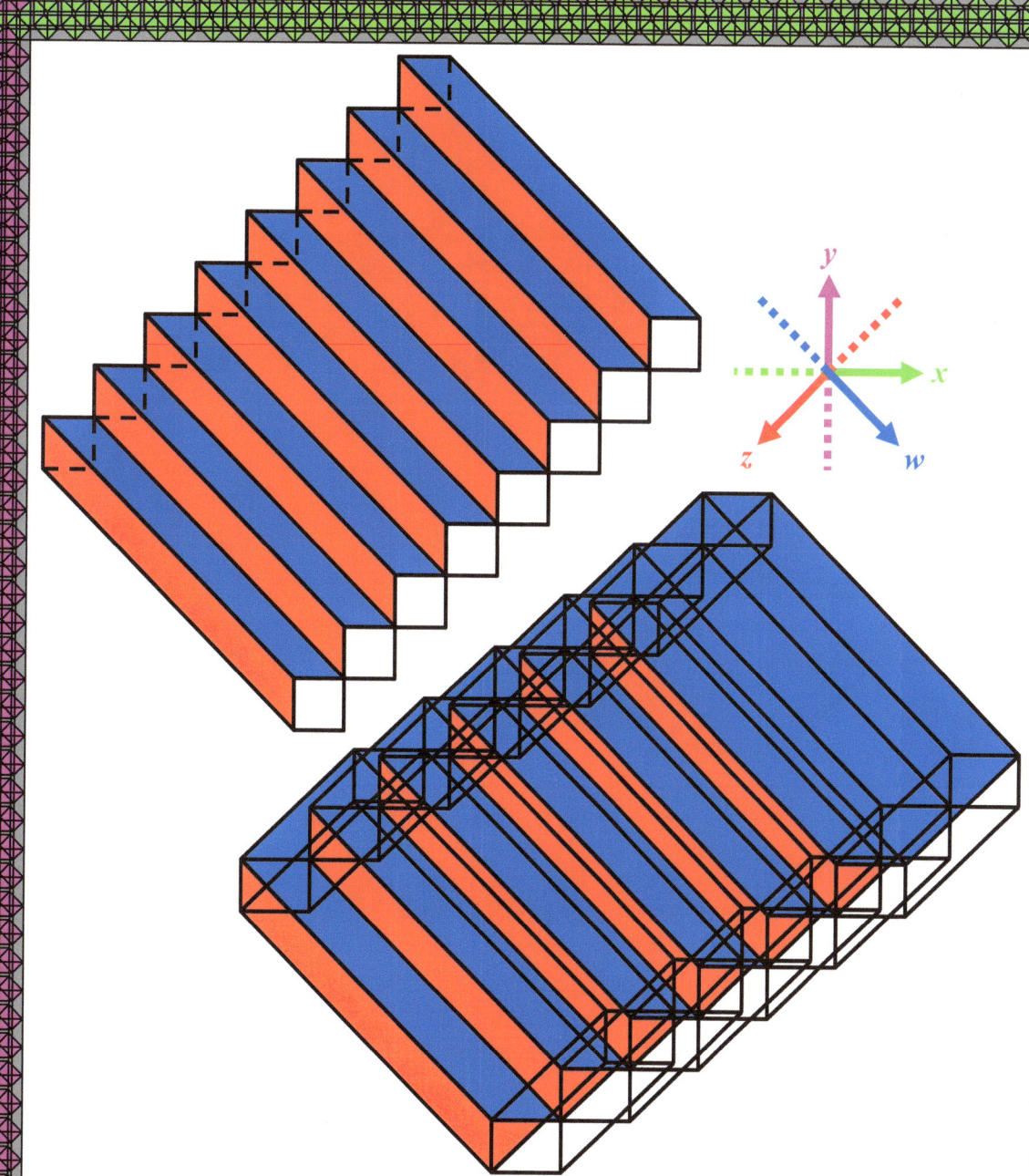

Compare the four-dimensional **hyperstaircase** on the right to the three-dimensional staircase on the left (in these diagrams, x, y, and w are the three common dimensions, while z is the fourth dimension). Either staircase provides a means to elevate a being upward along y, while also translating the being forward along x. In the three-dimensional staircase, two perpendicular rectangles in the wx and wy planes share adjacent edges; in the four-dimensional staircase, two orthogonal cuboids in the yzw and zwx hyperplanes share adjacent rectangles in the zw plane (the red cuboids are largely covered by the blue cuboids in the illustration).

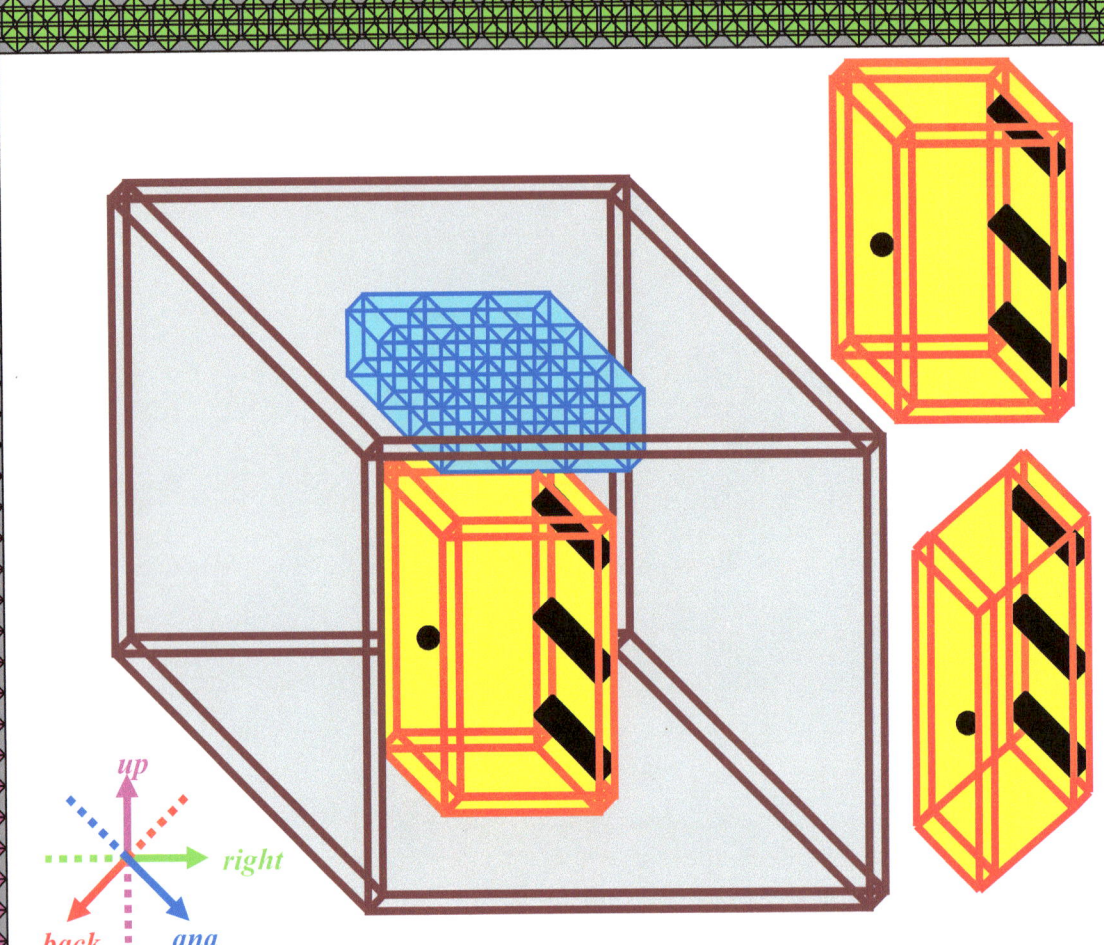

It's important to realize that the image above is merely one **hyperwall** – it is not a hyperroom. A hyperwall has three large dimensions in the shape of a cuboid (rectangular box), plus a small hyperthickness. This hyperwall is the front hyperwall of a hyperroom. The **hyperdoor** and **hyperwindows** are centered left/right and ana/kata; the hyperdoor extends upward from the bottom, but the hyperwindows do not reach the ceiling (they lie in the interior of the hyperwall). Looking from the left/right or ana/kata directions, or even looking down from above the ceiling, neither the door nor hyperwindows would be visible: They can only be seen from the forward/backward direction (and in this case a hyperbeing could see through the windows, and the door, if open). The glome-shaped **hyperdoorknob** lies in the interior of the hyperdoor, toward one side. The hyperdoor hinges at a rectangle (not at a linear edge like a usual door in three dimensions). The diagrams on the right side show the door in the closed (top) and open (bottom) positions. Compare to a typical wall in three-dimensional space, which is rectangular rather than cuboidal, has a rectangular rather than cuboidal door, and has square rather than cubic window panes; the windows of a rectangular wall in three dimensions could similarly not be seen from directions within the plane of the window, but only from the third dimension.

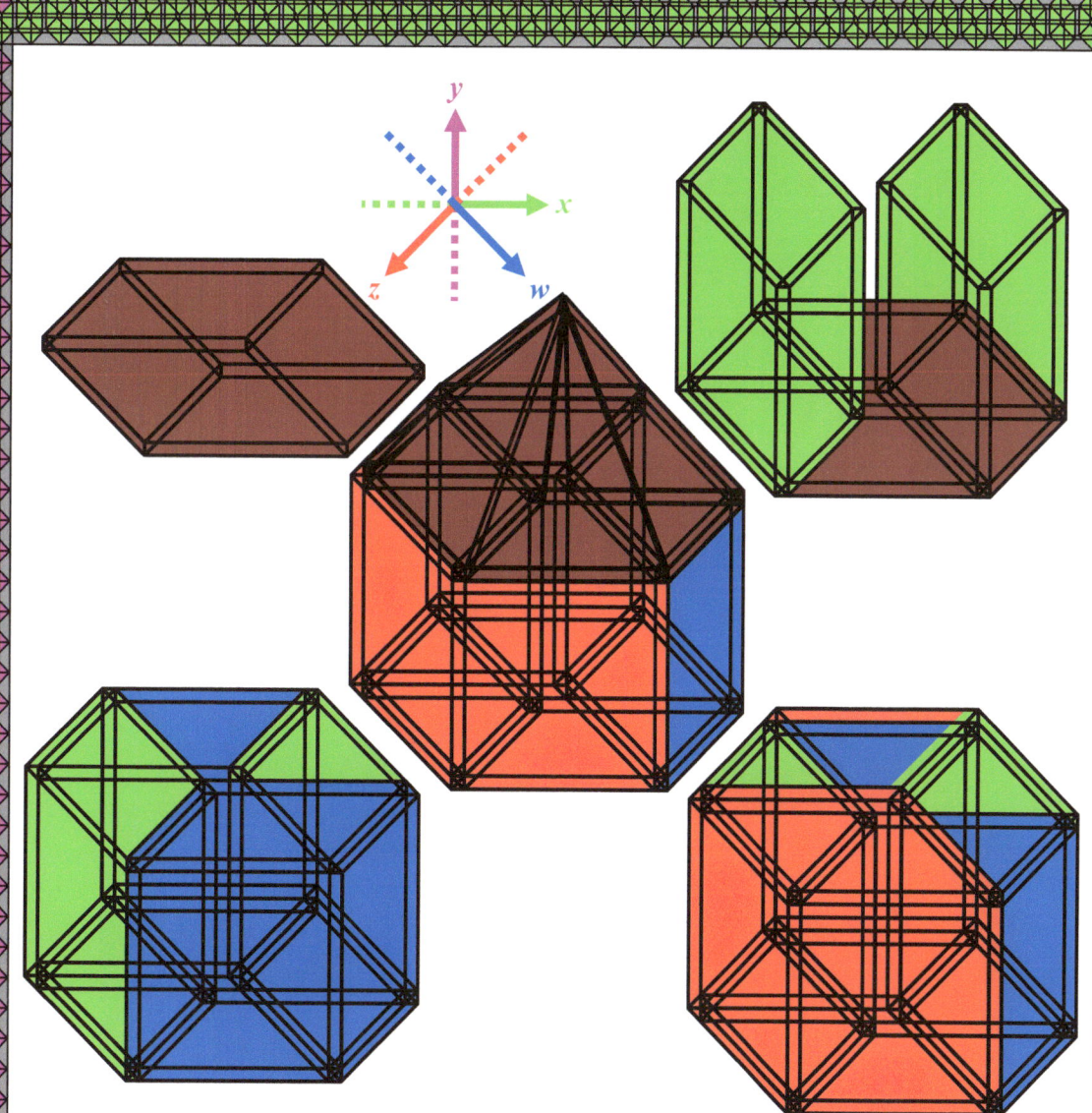

The illustrations above show a simple method of constructing a **hyperhouse**. The floor (top left) is cuboidal – three large dimensions (z, w, x) forming a cuboid (a rectangular box) and a small thickness along y. A pair of cuboidal walls (y, z, w) with small thickness along x are raised on the left and right sides (green walls). Similarly, pairs of cuboidal walls (x, y, z) with small thickness along w are raised on the ana and kata sides (blue walls) and pairs of cuboidal walls (w, x, y) with small thickness along z are raised on the front and back sides (red walls). Finally, a **hyperpyramid-shaped roof** with a small thickness along y provides shelter. All in all, there are 6 vertical walls, one floor, and one roof – corresponding to the 8 bounding cubes of a tesseract. Compare to a simple three-dimensional home shaped like a rectangular box, which has 4 vertical walls, one floor, and one roof, corresponding to the 6 bounding faces of a cube.

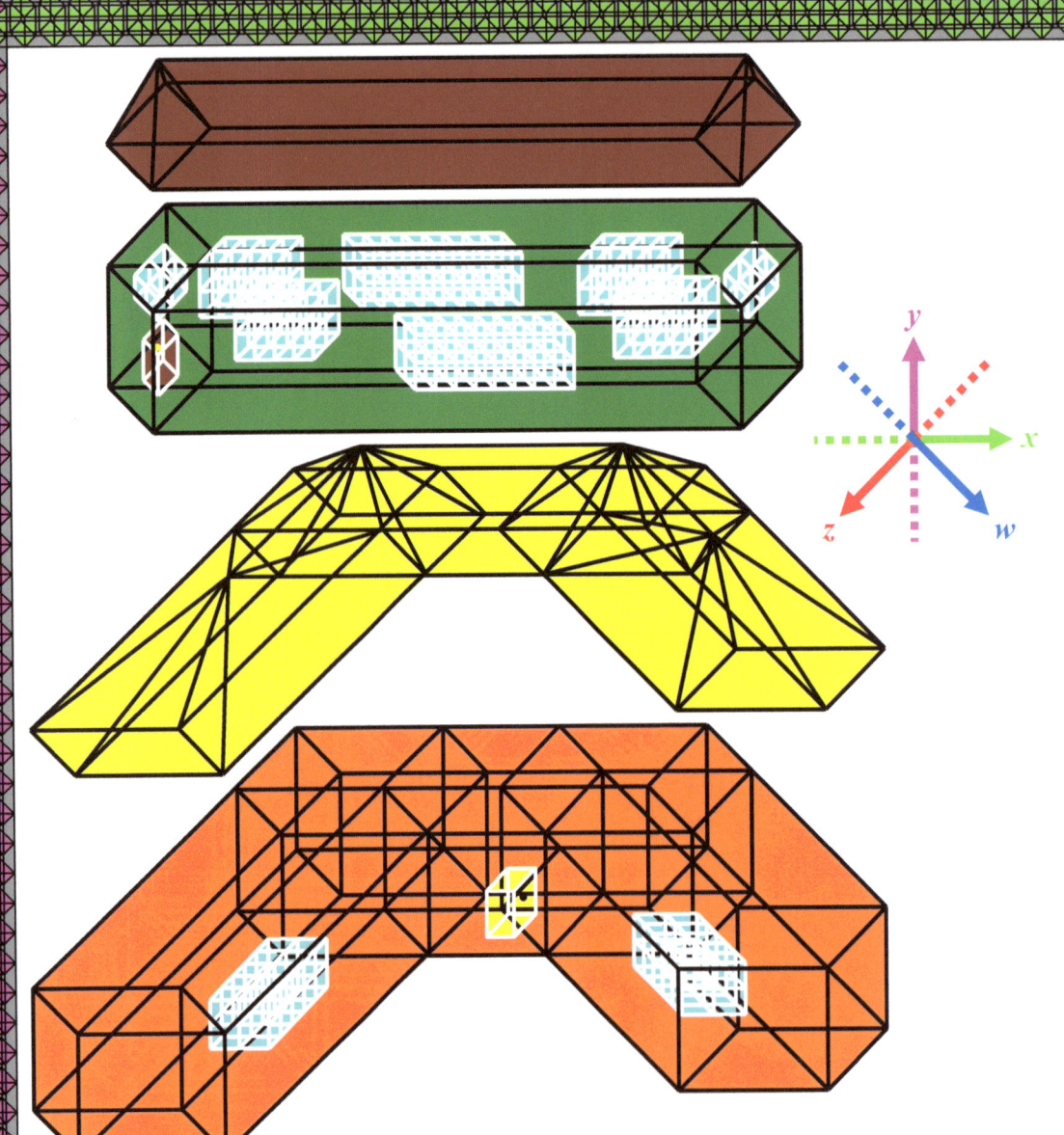

The top **hyperhome** is bounded by 6 effectively three-dimensional walls. Bear in mind that each of these three-dimensional walls is very thin compared to the four dimensions of the hyperhouse: That is, the four-dimensional living space inside the hyperhome is enormous compared to its thin hyperwalls. Similarly, the hyperdoor and hyperwindows are tiny and only found at the three-dimensional boundaries of the four-dimensional hyperhome. Three different types of hyperwindows are shown – corresponding to front/back, left/right, and ana/kata. The hyperroof is illustrated above the hyperhouse – it fits snugly right on top – to offer a better view of the living space. The bottom hyperhome has a more complicated floor plan and roof structure. Only a couple of sets of hyperwindows are drawn to prevent the image from becoming too busy.

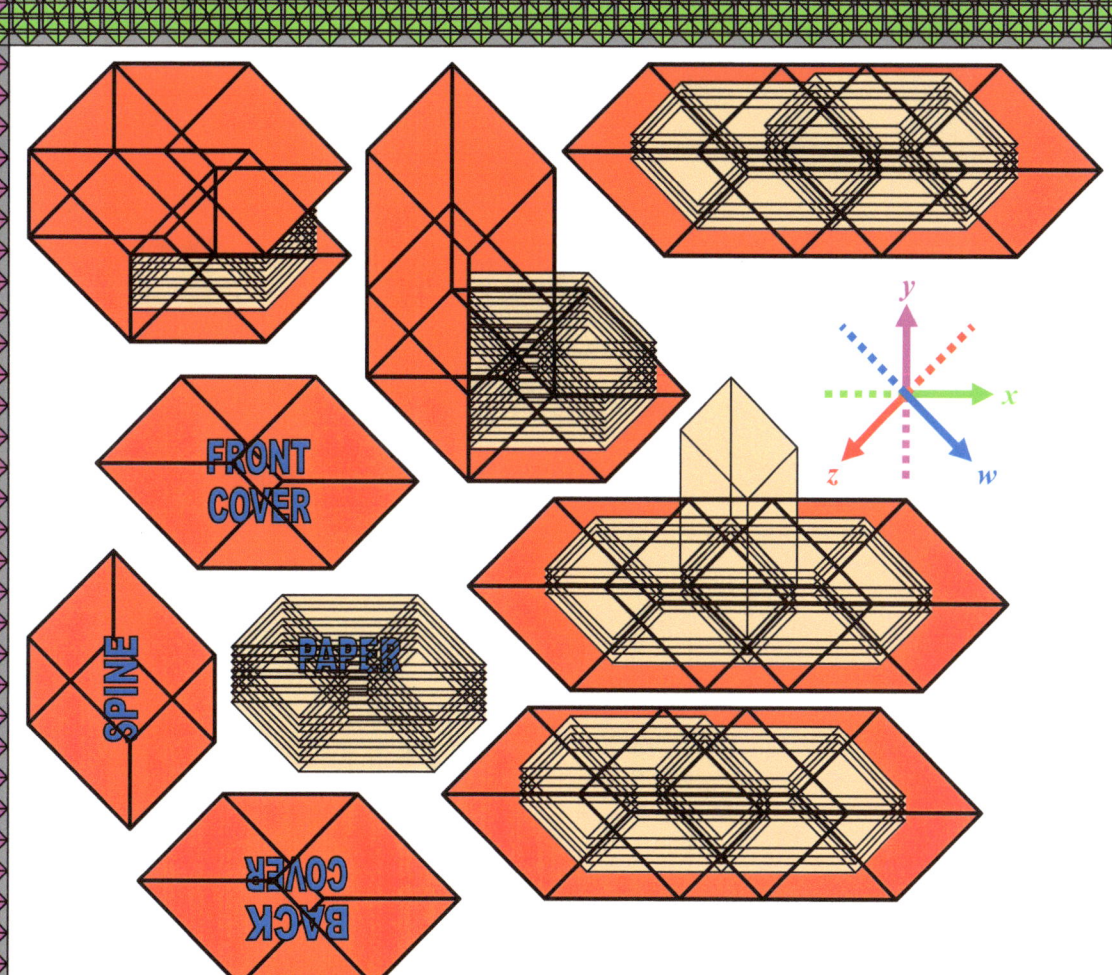

In the fourth dimension, **hyperpaper** would be three-dimensional, shaped like a cuboid, with a small hyperthickness. A stack of hyperpaper would essentially be a stack of three-dimensional rectangular boxes, where each point in the interior of one box would be touching the corresponding point of the interior of adjacent boxes. Compare to three-dimensional space, where rectangular boxes can only share at most a rectangular face. Stacking cuboids is almost as difficult to imagine as it would be for a two-dimensional being to imagine stacking rectangles in three dimensions. The top left image shows a closed **hyperbook**, which has essentially a three-dimensional cuboid-shaped front cover, spine, and back cover. One rectangular side of each hyperpage meets at the spine (cf. three dimensions where page edges meet at the spine). Three-dimensional diagrams and text could be written on the front and back 'sides' of each hyperpage, and the entire three-dimensional image on either side (front or back) could be seen without any problems gauging depth. Instead, illustrators and readers would face the challenge of gauging **hyperdepth** (the fourth dimension). The other diagrams above show how to open the book and turn the pages. Each three-dimensional hyperpage reflects upon a 180º rotation as the hyperpage is turned. One may read left to right, ana to kata, and top to bottom.

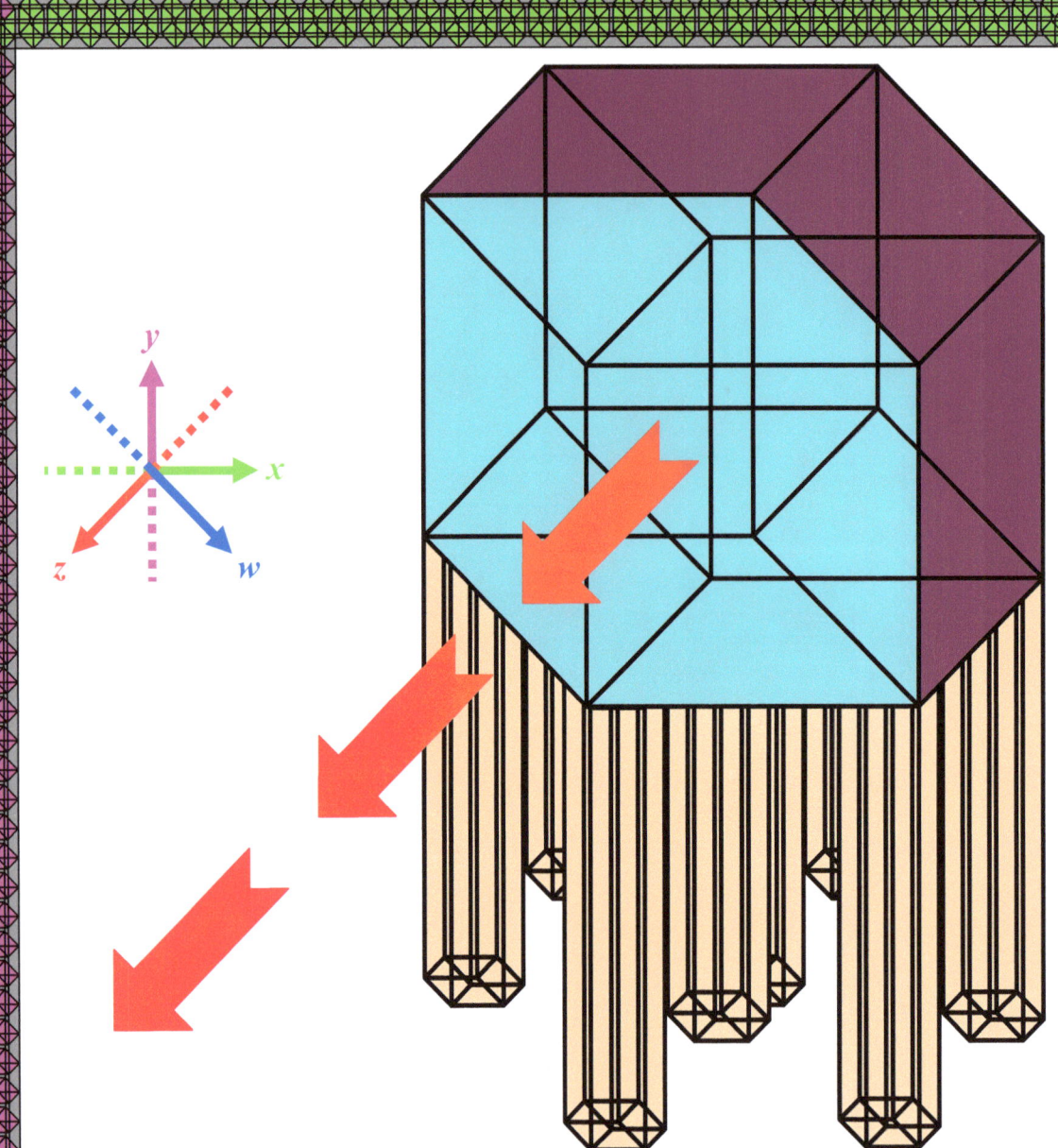

This **hypertelevision** features the shape of a **hypercuboid** (like a tesseract, but not necessarily equal lengths of edge types). One of the 8 bounding cubes serves as the **hyperscreen** (wxy). Note that this is a flat hyperscreen hypertelevision, but not slimline: That is, the hyperscreen is flat (not curved), but it is not slim because it has significant depth along the fourth dimension z. Hyperbeings would sit on a hypercouch and view this hypertelevision from the positive z-axis. They would observe three-dimensional pictures – they could see every point on (they would say *on*, not *in*) the cube-shaped hyperscreen. Hyperstereo would feature sound from four directions – front left, front right, front ana, and front kata; surround sound would have four additional speakers – back left, back right, back ana, and back kata.

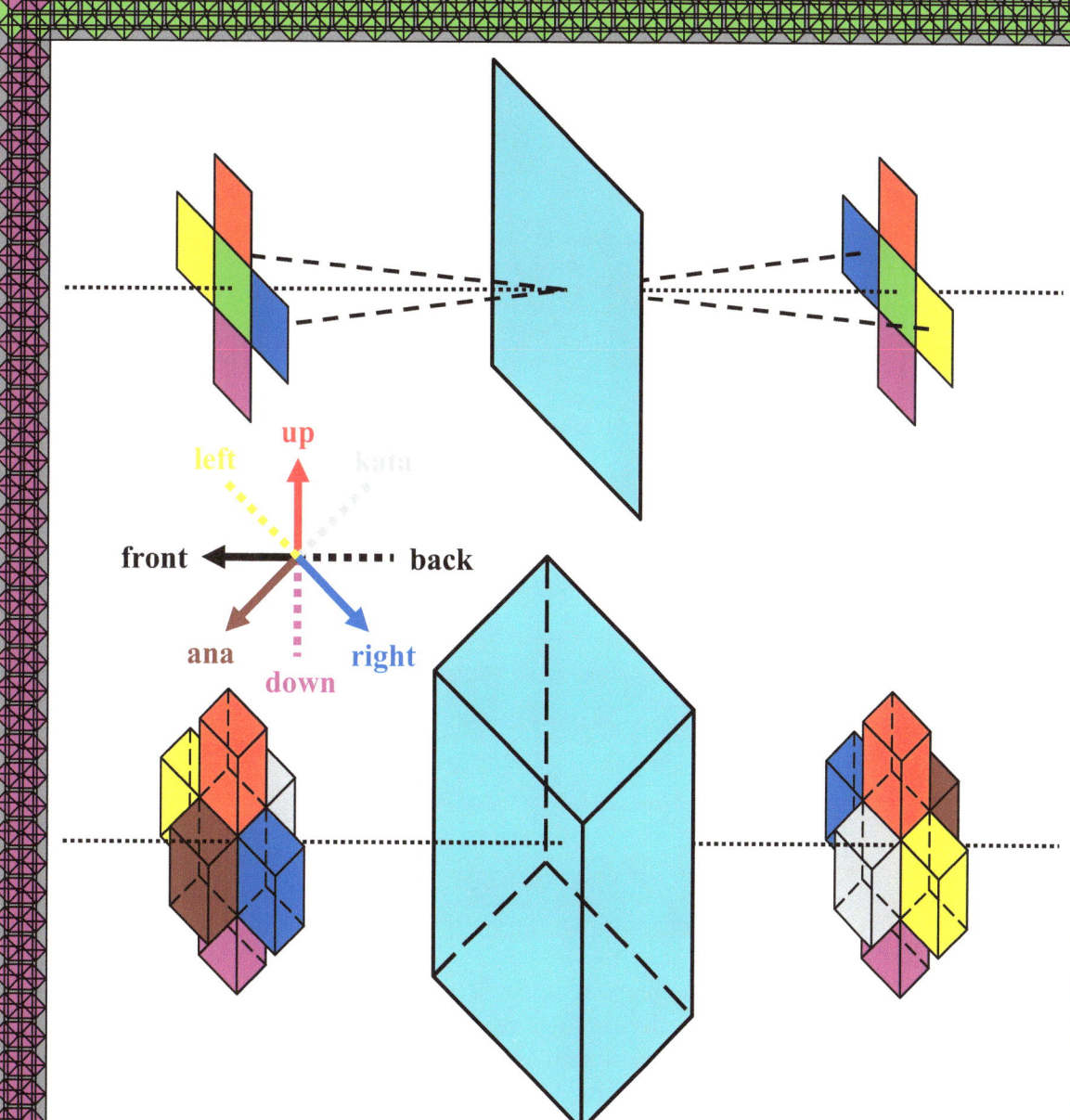

The top illustration shows a two-dimensional object's reflection in the third dimension when viewed through a plane mirror. Note that the original object cannot be rotated into its reflection within its plane, but can be rotated through the third dimension about the vertical axis to look like its presently reflected image. Similarly, a three-dimensional object is placed before a **hyperplane hypermirror** in the bottom illustration. **Four-dimensional hyperbeings** with left, right, ana, and kata **hypereyes** would perceive reflections of ana/kata in addition to left/right. This object cannot be rotated to match its image within its three-dimensional hyperplane, but can be rotated through the fourth dimension about a plane with one vertical dimension to attain the same configuration as its current image.

Part II: Hyperspherical 4D Objects
(Glome-Based Objects)

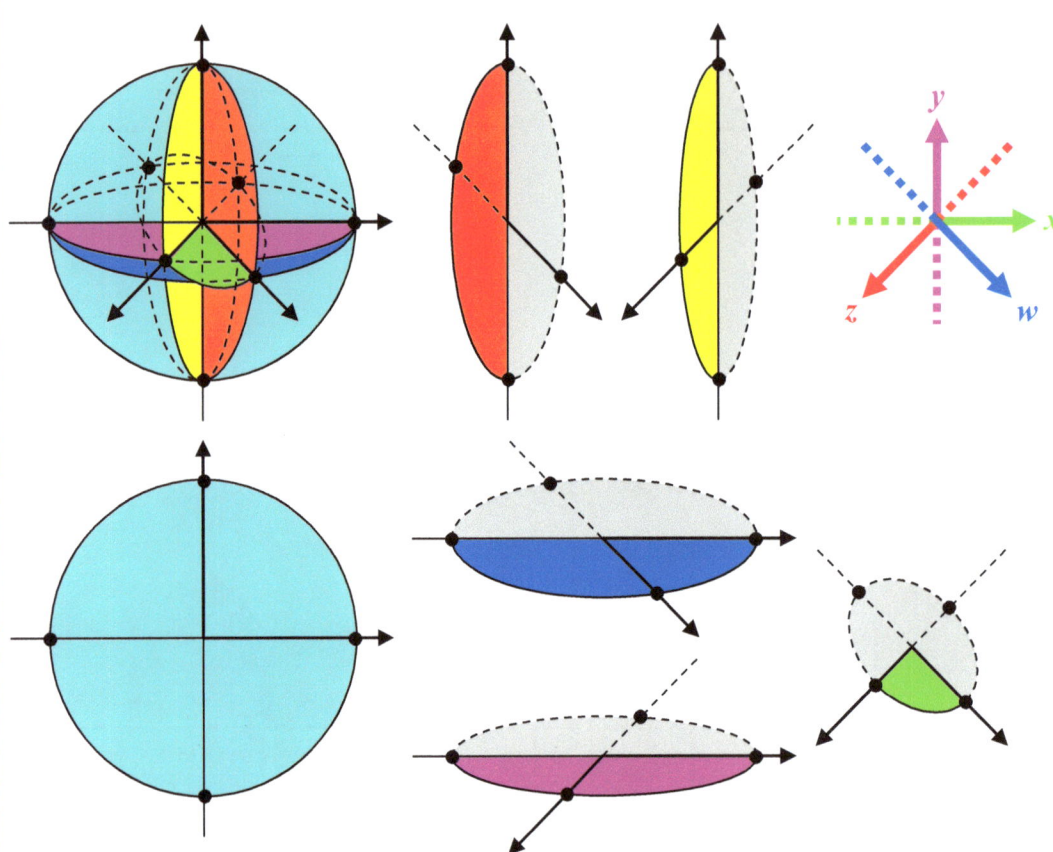

The **glome** is a **hypersphere** in four-dimensional space – counting the dimensions with a Euclidean mindset, for the **hypersurface** of the glome has three independent dimensions. The glome refers to just the hypersurface, whereas the similar solid object that includes the interior is termed a **hyperball**. This glome (top left) is illustrated by a set of 6 mutually orthogonal **great circles** (a great circle is a circle with its radius equal to the radius of the glome – like a longitude, or like the equator, but not like the other latitudes) – one great circle in each of 6 mutually orthogonal planes (xy, yz, zx, wx, wy, and zw). Only the frontmost ($z > 0$) and anamost ($w > 0$) sections of the great circles are colored in the glome, in order to improve their visibility; while the rear and kata sections are depicted with dashed curves. As with the tesseracts that have been illustrated in this volume, the glomes are shown with greater hyperdepth than depth (i.e. the glome appears to extend further along w than along z, which is merely a perspective illusion since the glome naturally extends equal proportions in any direction in four-dimensional space). The purpose of this is to enhance features that would otherwise be obscured in symmetric geometric objects.

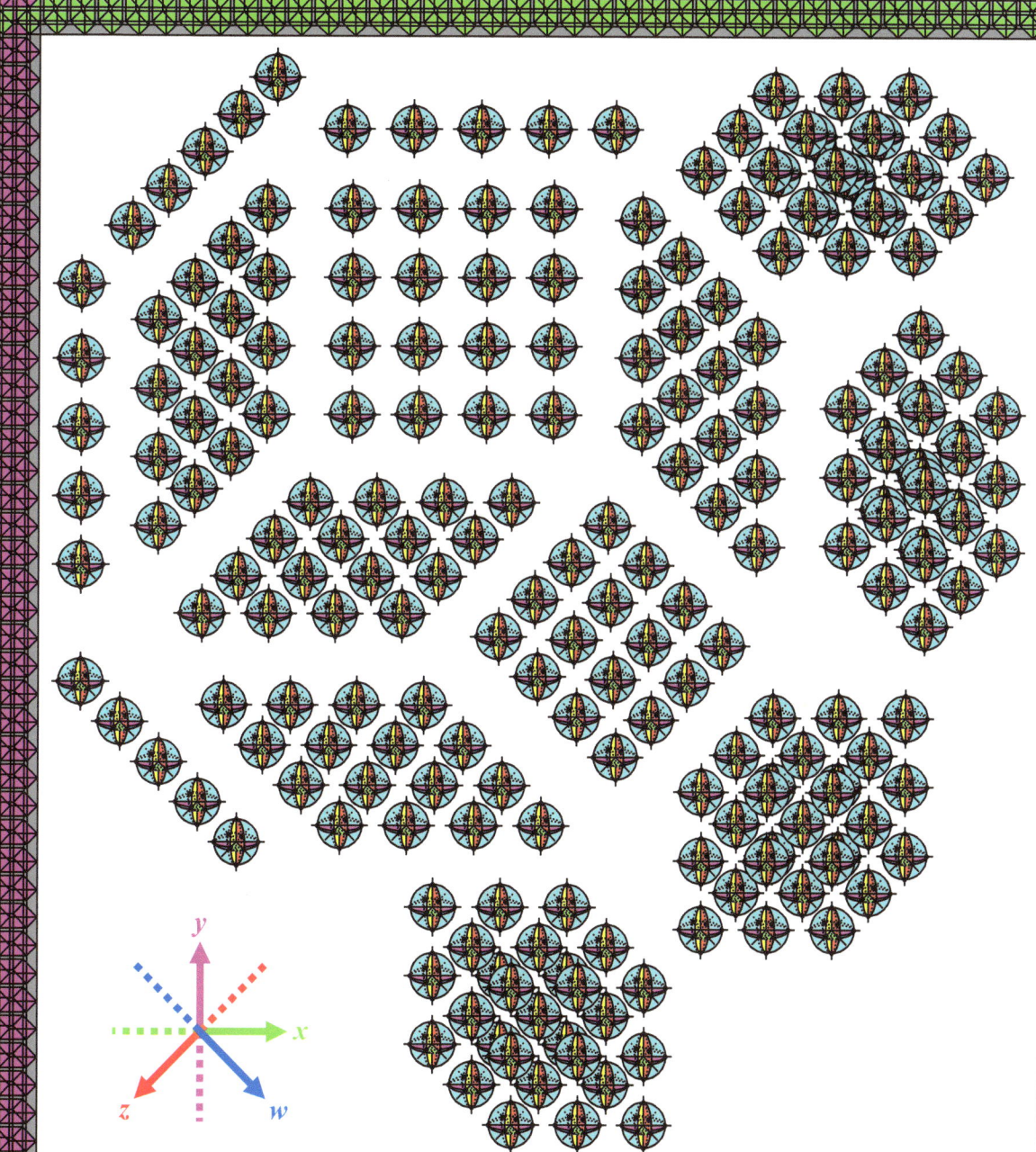

This illustration shows the fundamental one-, two-, and three-dimensional **arrays of glomes**. The 4 linear chains – one along each of the coordinate axes x, y, z, and w – appear on the top and left edges of the diagram. The 6 planar configurations – one lying in each of the mutually orthogonal planes xy, yz, zw, wx, zx, and wy – occupy the upper central portion of the picture. The 4 hyperplanar configurations – one lying in each of the mutually orthogonal hyperplanes xyz, yzw, zwx, and wxy – are located on the right and bottom edges of the illustration. These basic building blocks can be stacked together to form higher-dimensional configurations.

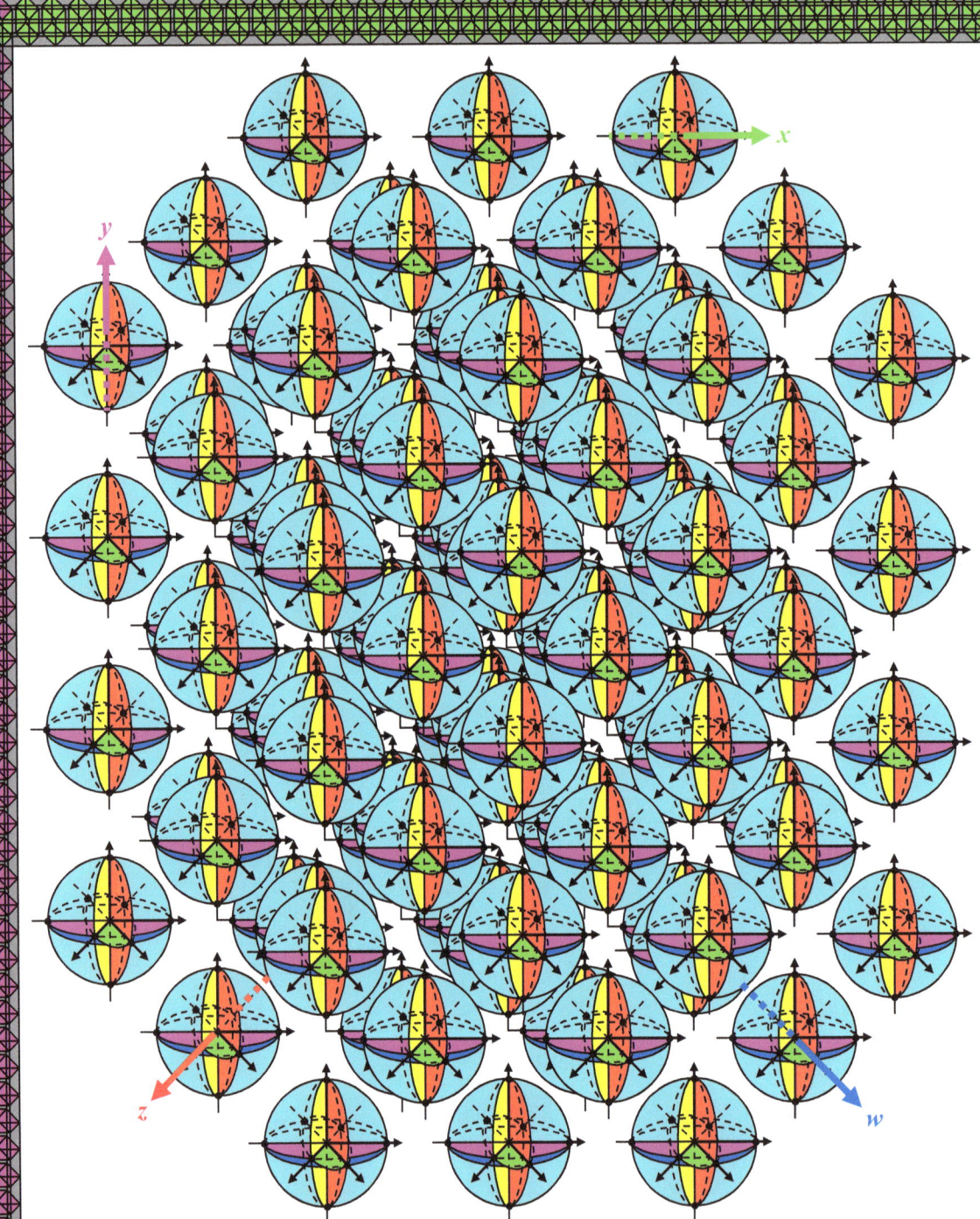

This **simple hypercubic lattice** is the most basic **four-dimensional array**. It is built up from unit tesseracts with a glome on every corner. The section illustrated above is 3 glomes wide, 4 glomes high, 3 glomes deep, and 3 glomes hyperdeep. Many **hypercrystals** in the fourth dimension would feature such a lattice structure, with atoms, ions, atom groups, or molecules at each corner. A four-dimensional **hypercheckerboard** is based on a similar structure.

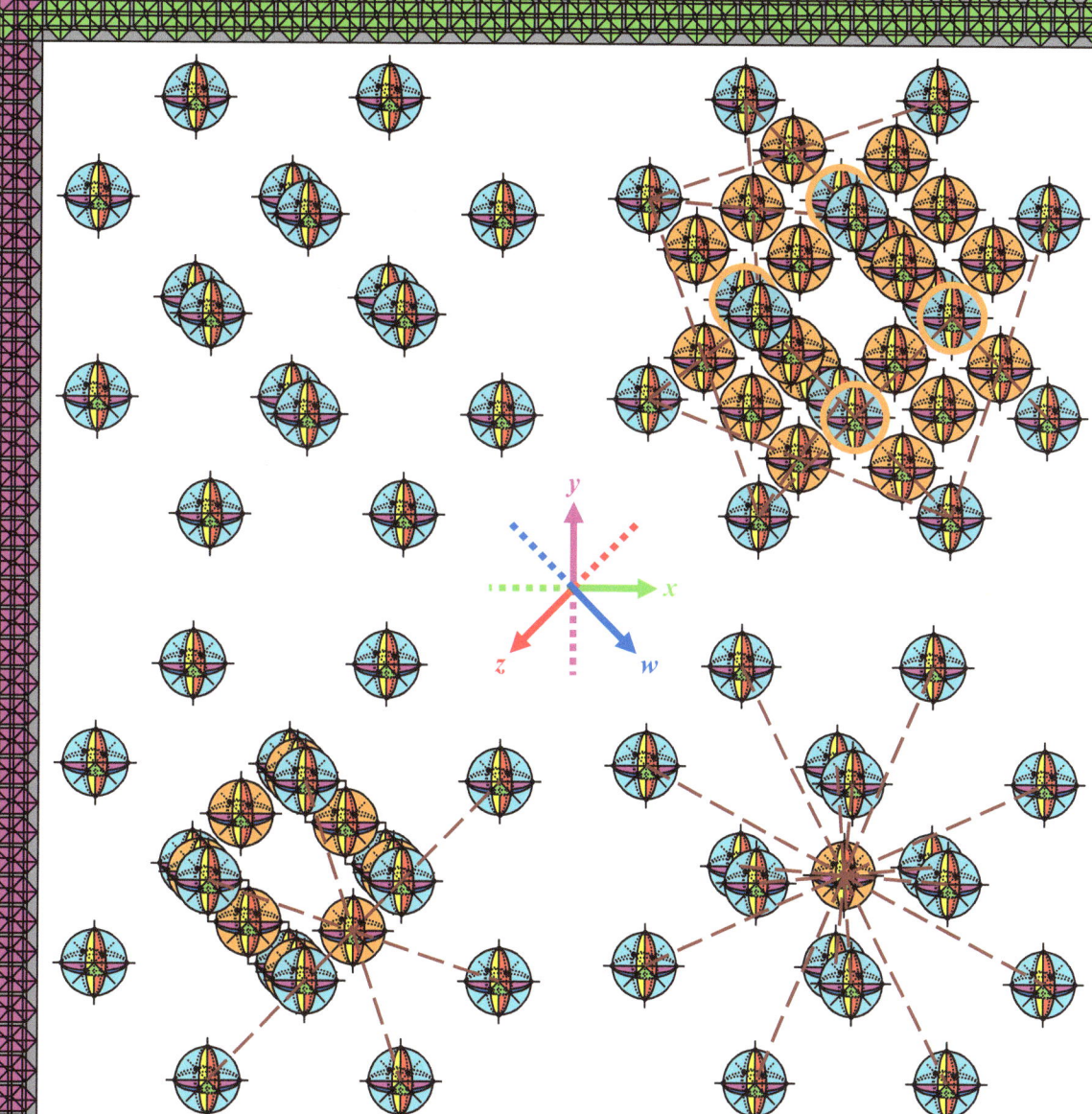

Four fundamental types of **hypercrystals** are illustrated above. Each diagram shows one unit cell with turquoise-colored glomes at the 16 corners of the unit tesseract. The **simple hypercubic lattice** unit cell is shown at the top left. The **content-centered hypercubic lattice** appearing on the bottom right has one additional **brown**-colored glome at the center of the unit tesseract. The **cube-centered hypercubic lattice** positioned on the bottom left has 8 **brown**-colored glomes in addition to the 16 turquoise-colored glomes in its unit cell – one at the center of each of the 8 bounding cubes. The square-centered hypercubic lattice shown at the top right has 24 **brown**-colored glomes in addition to the 16 turquoise-colored glomes in its unit cell – one at the center of each of the 24 bounding squares comprising the 8 bounding cubes (4 for each of the 6 types of mutually orthogonal squares); in this diagram, 4 **brown**-colored glomes that would otherwise be hidden appear as outlines.

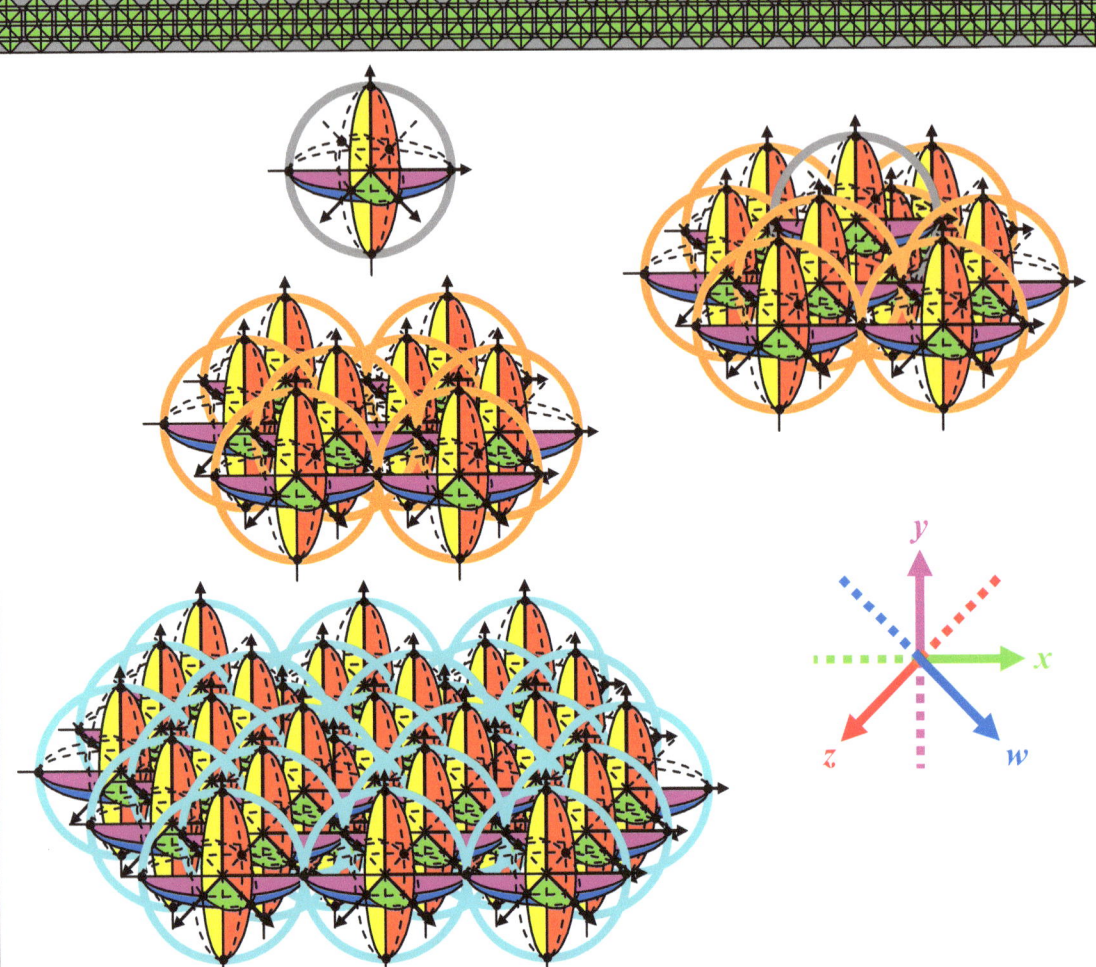

The **content-centered hypercubic lattice structure** is the efficient method of packing that would be employed to stack glome-shaped **hyperfruits**, like **hyperoranges**, at a **hypersupermarket**. Starting at the bottom of a **hyperbox**, the hyperfruits would be stacked in an $N \times N \times N$ (three-dimensional) simple cubic lattice (or $N \times M \times L$ if the base is cuboidal rather than cubic). On the second row (the next row up), the **hypergrocery store stockperson** would position hyperfruits directly above the center of each unit cube of hyperfruits on the first row. The third row would repeat the first row, completing one layer of content-centered hypercubic lattice – i.e. each hyperfruit on the second row lies at the center of a unit tesseract. Each even row would be identical to the second row and each odd row would be identical to the first row until reaching the top of the box. At the top of the box, the pattern would be the same, except that it would become hyperpyramid shaped – to prevent hyperfruits from falling over the edges. The hyperpyramid structure at the top is illustrated in the diagrams above. The top three layers are shown. At the very top is a single hyperfruit, the second row contains 8 hyperfruits, the third row has 27 hyperfruits, and the Nth row has N^3 hyperfruits (assuming a cubic base). The layers are vertically separated on the left for easy visibility; the top two layers are joined at the right.

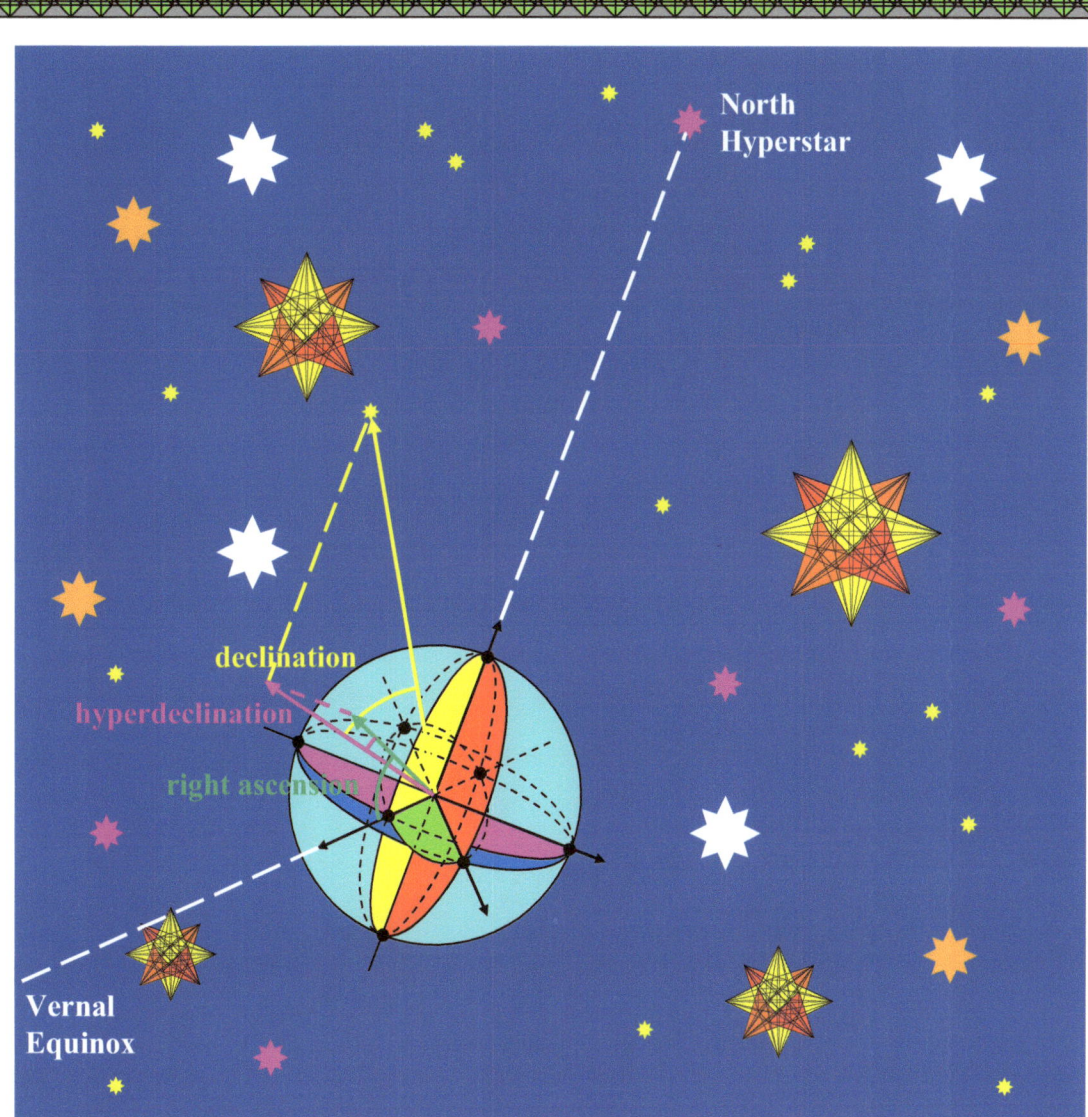

The location of a **hyperstar** in the night sky in four-dimensional space as viewed from a spinning **hyperplanet** could be specified in terms of **celestial hypercoordinates**. Imagine a vector pointing from the hyperplanet to the hyperstar. The **declination** would measure the angle between this vector and its projection onto the **celestial equatorial sphere**; the **hyperdeclination** would then be the angle between this vector's projection onto the celestial equatorial sphere and its subsequent projection onto the **celestial equator**; and the right ascension would measure this final projection's orientation relative to **vernal equinox**. This celestial hypercoordinate system is very similar to **hyperspherical coordinates**. Declination equates to **latitude** on the **celestial glome**, hyperdeclination corresponds to **hyperlatitude**, and right ascension is a **celestial longitude**. A **north hyperstar** would define the north celestial pole – it would be analogous to Polaris for present-day earth observers.

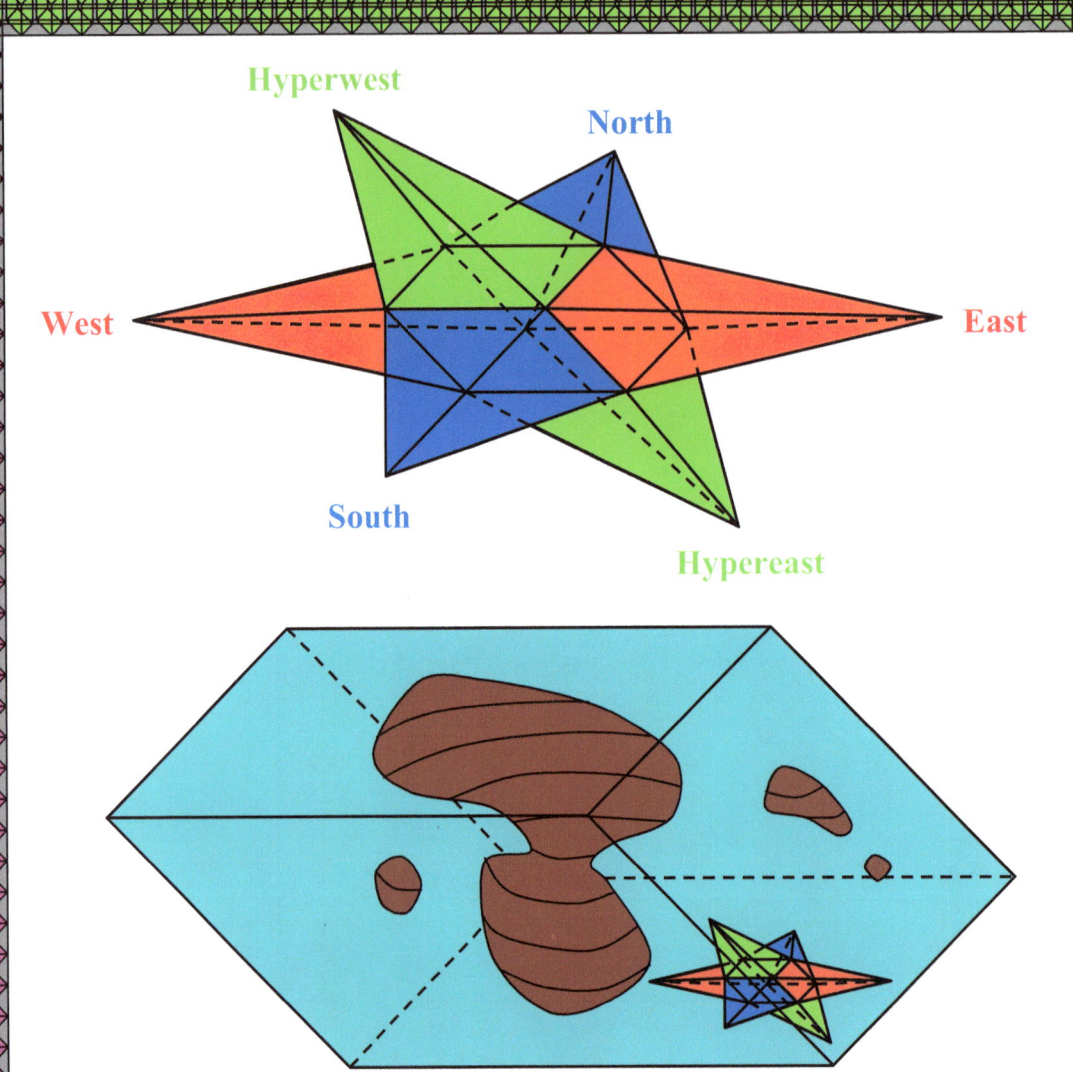

Four-dimensional **hyperbeings** living on the **hypersurface** of a **glome-shaped hyperplanet** would use a **hypercompass** with 6 perpendicular directions: North/south, east/west, and **hypereast/hyperwest**. The north end of the **hypercompass needle** would be attracted to the magnetic south pole of the planet (near the geographic north pole, if the naming scheme parallels English geography on earth). East and west would point to the hyperplanet's rising and setting **hypersun**, respectively. While the hypersun would appear to travel along a **hyperecliptic**, which defines an **equatorial circle**, the hyperplanet would actually have an **equatorial sphere** – since there are two dimensions, east and hypereast, that are perpendicular to north and to each other. The bottom diagram above shows a simple three-dimensional **hypermap** drawn on a three-dimensional **hyperpaper**. Hyperbeings could see and draw anywhere *on* the three dimensions of the hypermap. Hyperbeings could see three-dimensional **hypercontinents**, while the fourth dimension (altitude) would have to be depicted via topological surfaces (__not__ depicted on the illustration above).

The illustration above shows four snapshots of a **hyperplanet** as it revolves around its **hypersun**, while also rotating about a **tilted plane** (cf. tilted axis in three-dimensional space). In the four positions shown, this hyperplanet rotates one-quarter turn about its planar axis, but travels less than a quarter revolution around its hypersun. This hyperplanet rotates about an axis midway between the $w'x'$ and $w'y'$ planes as it revolves about its hypersun in the zx plane. The **spin coordinates** (primed) are tilted relative to the **revolution coordinates** (unprimed). The y' axis points to the North Hyperstar at the **celestial north pole**. During the rotation shown, three of the colored great circles' projections onto the plane of this page grow as the other three shrink, while the orientation of the planar axis of rotation is preserved. While the hypersun would be glome-shaped, the one illustrated above is artistically depicted as a **hyperstar-shaped polytope** – a **hyperpyramid** attached to each cube of a tesseract.

Hyperintelligent hyperbeings who appreciate their rich geometry may adopt a number system (top) and alphabet (bottom) like those illustrated above. The number 1 is a plane; numbers 2 and 3 are as many intersecting planes; and numbers 4 thru 8 are polyhedra with as many sides. Thinking three-dimensionally – like the symbols they could write on **hyperpaper** – these basic 8 digits can be arranged in a cubic array. Such hyperbeings could have two sets of opposable **hyperfingers**, of four and three, and one mutually opposable **hyperthumb** – to match this geometric number system. The dash and sphere naturally represent zero and infinity. The samples in green translate as 42 and 3.14. The alphabet has 27 letters, which can also be arranged in a cubic array. The center letter in green represents the usual vowel sounds – an accent mark indicates which. The 6 square-centered letters are other vowel or vowel-like sounds (w, y, h, r, oi, ow). The 12 edge-centered letters are consonants that require use or positioning of the **hypertongue** (like d or l); this also includes sounds like th, *th*, sh, and ch. The remaining 8 consonant sounds appear on corners (like b or m). They may organize the letters based on where/how the **hypermouth** pronounces them. Such hyperbeings might write *the fourth dimension* as shown at the bottom.

This sample **hyperbilliards** game illustrates the three-dimensional **hypersurface** of the **hyperpool table**. The **glome-shaped hyperballs** roll along the three-dimensional **hypertabletop**: They can roll right/left, ana/kata, or front/back without moving vertically. This hyperpool table has 26 **hyperpockets** – 8 corners, 12 edges, and 6 sides. The hyperballs are arranged with four square arrays stacked in the right/left direction, nestled together. There are 30 hyperballs: $1^2 + 2^2 + 3^2 + 4^2 = 30$. The **hyper-cue-stick** would be **spherindrical** (a **spherinder** is a **hypercylinder** with spherical cross section). The hyperballs could spin right, left, ana, kata, forward, backward, or combinations of this, allowing for rich possibilities of **hyper-English**.

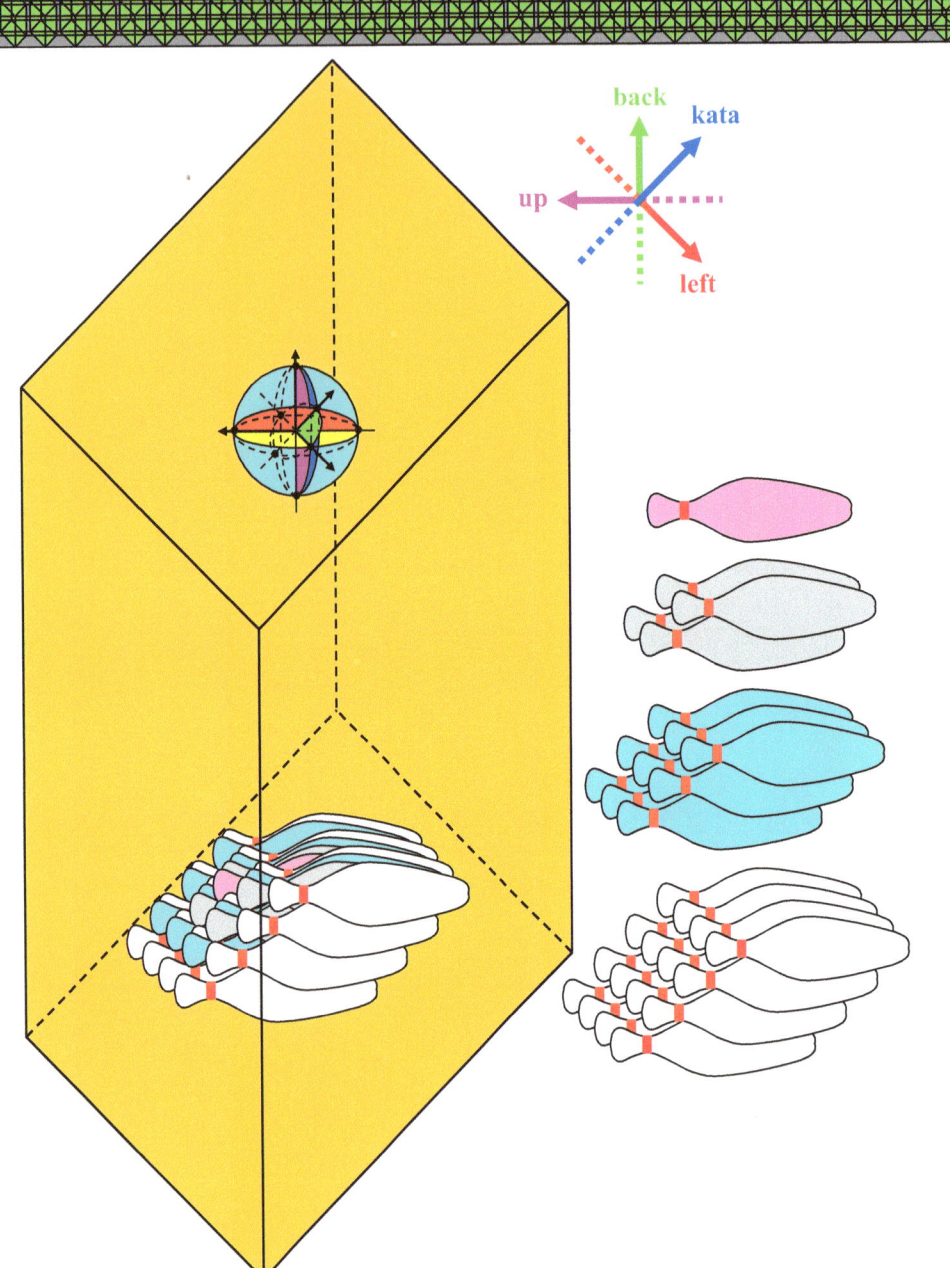

Hyperbowling has features similar to **hyperbilliards**. The **hyperbowling hyperball** is glome-shaped: It would probably have four or five holes for **hyper-humanoid beings**, which would be well-adapted to have either two opposable **hyperthumbs** and one set of **hyperfingers** or one opposable hyperthumb and two sets of opposing hyperfingers. There are 30 **hyperpins** arranged in four planar 'rows.' As with hyperbilliards, only the spherical (not circular) base of each hyperpin touches the three-dimensional hypersurface of the bowling lane; the bulk of each hyperpin lies above the lane (though it may not appear that way in the two-dimensional projection above). Similarly, the pink hyperpin is actually at the front of the arrangement.

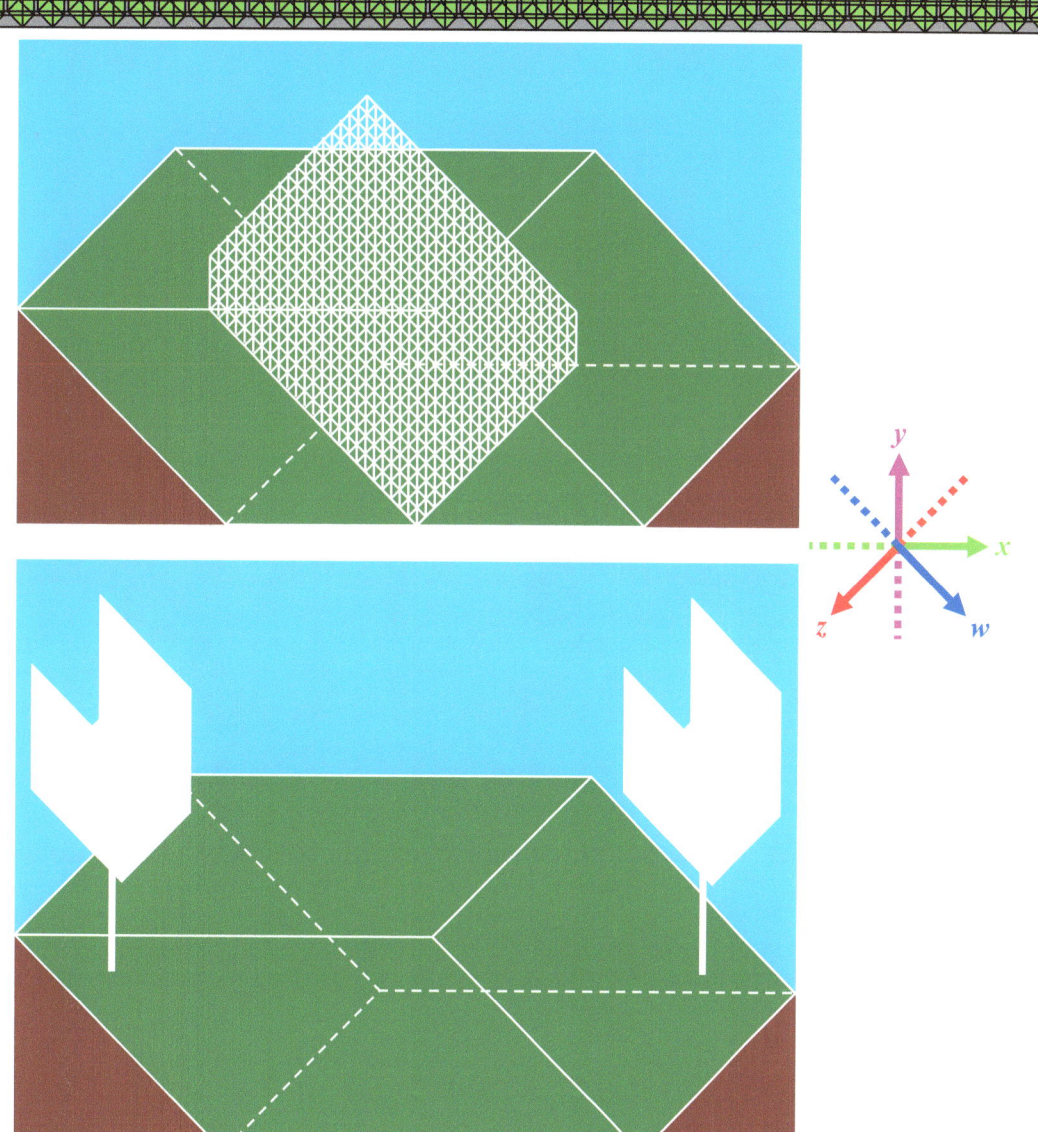

In the fourth dimension, the **hyperground** would be a three-dimensional hypersurface. **Hyperathletes'** **hypershoes** would have three-dimensional **hypersoles**. Hyperathletes could run forward/backward ($\pm x$), right/left ($\pm z$), or ana/kata ($\pm w$) without moving vertically. The top diagram illustrates a **hypertennis court**, which features a three-dimensional **hypernet** with a few feet of height (but negligible hyperthickness in its fourth dimension). The frame of a **hypertennis racket** would be spherical and the strings would be strung in a **criss-cross-hypercross** pattern. The spherical racket head would still be flat in the fourth dimension. Service planes (cf. lines) would be painted on the court. The **hyperfootball field** in the bottom diagram resembles the hypertennis court. Five and ten yard planes would be painted on the grass. A hyperfootball field goal kicker would strive to kick a hyperfootball through the three-dimensional **hypergoal**, which has only minute thickness along x.

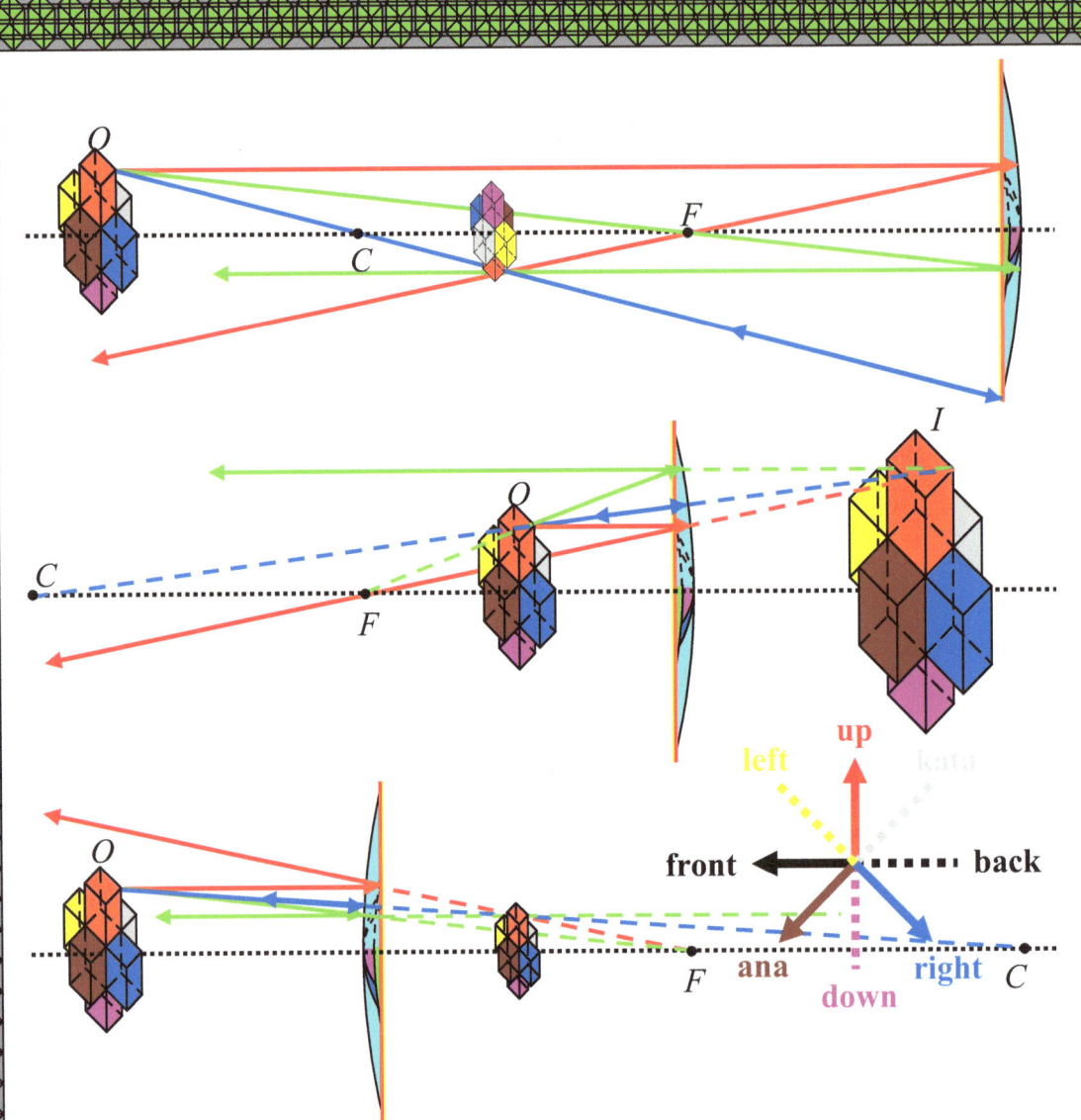

These diagrams illustrate the reflection of a three-dimensional object placed before a **hyperspherical hypermirror** in four-dimensional space. The top two diagrams involve a concave hypermirror, while the bottom hypermirror is convex. Image formation follows the same ray tracing rules as in three dimensions. The main difference is that inverted images are also inverted in the ana/kata sense. A truly inverted image, such as in the top diagram, inverts top/bottom, right/left, and ana/kata; the bottom images are upright. A **hyperplane hypermirror** (cf. p. 27) actually has an upright image – it is only perception that is inverted. As in three dimensions, a plane mirror does not turn a west hand into an east hand – it just gives the perception that the hand on the right side is a left hand. When we view objects with mirrors, we see them from a different perspective, which causes words help up before mirrors to appear backwards. Yet east is still east, not west, so we term the image upright for a plane mirror.